C0-APY-476

A Practical Guide
To
Building Construction

A PRACTICAL GUIDE
TO
BUILDING CONSTRUCTION

EDGAR LION

Professional Engineer

PRENTICE-HALL, INC., ENGLEWOOD CLIFFS, N.J. 07632

Library of Congress Cataloging in Publication Data

Lion, Edgar,
    A practical guide to building construction.

    Includes index.
    1. Building. I. Title.
TH145.L48   690   80-10141
ISBN 0-13-690628-1

Editorial/production supervision and interior design by
Nancy Moskowitz and Mary Carnis
Cover Design by Edsal Enterprises
Manufacturing Buyers: Gordon Osbourne and Anthony Caruso

© 1980 by Prentice-Hall, Inc., Englewood Cliffs, N.J. 07632

All rights reserved. No part of this book
may be reproduced in any form or
by any means without permission in writing
from the publisher.

Printed in the United States of America

10 9 8 7 6 5 4 3 2 1

Prentice-Hall International, Inc., *London*
Prentice-Hall of Australia Pty. Limited, *Sydney*
Prentice-Hall of Canada, Ltd., *Toronto*
Prentice-Hall of India Private Limited, *New Delhi*
Prentice-Hall of Japan, Inc., *Tokyo*
Prentice-Hall of Southeast Asia Pte. Ltd., *Singapore*
Whitehall Books Limited, Wellington, *New Zealand*

Contents

4   CONTRACTING: PLANNING, METHODS AND CONTROLS   145

CHECKLISTS

Preface

Construction is not only one of our biggest industries, but because of its dynamic nature it is one of the most complex industries in our economy. Partly because of humanity's quest for progress, partly because of business competition, and partly because of the inventions and developments in other fields, construction has advanced in a manner our ancestors of a hundred years ago would probably not have believed possible. Because our global population is increasing on a geometric scale, the potential for construction in all its fields all over the world is enormous. By the same token, our dwindling resources and our socioeconomic problems have placed increasing stress on our ingenuity to try to improve the end product. New materials and products constantly appear on the commercial horizon, new equipment is designed, existing equipment is improved, and new construction methods are devised, sometimes through sheer necessity. Some of these innovations will remain, others will disappear, but certain basic materials, equipment, and methods will stay with us for a long time to come. The changes in all types of construction are proliferating at an enormous rate in all aspects, as witnessed by the wealth of technical literature and publications.

It is this superabundance of knowledge that motivated the creation of this book. There are many highly specialized technical books on the market dealing with all kinds of subjects, and there are a number of highly technical handbooks brimming with tables, statistics, and graphs. These books are excellent for their technical content, but, they usually ignore items which might fall into other categories, and, in addition, they force the reader to look through many pages to find even simple references. For the person, whether layman or expert, looking for information that is in simple and in concise form and that is easy to find, this book is not only the answer, but it is also a valuable reference book to be consulted many times for varied purposes. The information herein is intended to give the layman or person on the periphery of construction or in professions in related fields and the professional looking for easy references and aids in preparation of his work, a comprehensive general coverage of building construction from initial concept to final completion.

Edgar Lion, *Professional Engineer*

1

Introduction

CLASSIFICATIONS OF CONSTRUCTION

This book deals specifically with building construction, but in order to give the reader a better perspective and a general overview, construction is presented systematically separated into various divisions and subdivisions.

Because of its complex nature, construction can be broken down in different ways and there is no uniformly standardized system in that respect. The following classification, which follows fairly conventional usage, is a suggested format for a systematic approach.

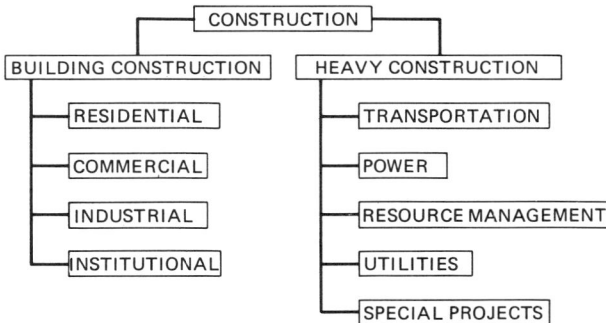

```
                    ┌─── CONSTRUCTION ───┐
       ┌────────────────────┐     ┌───────────────────┐
       │ BUILDING CONSTRUCTION │   │ HEAVY CONSTRUCTION │
       └────────────────────┘     └───────────────────┘
          ─┤ RESIDENTIAL │           ─┤ TRANSPORTATION │
          ─┤ COMMERCIAL │            ─┤ POWER │
          ─┤ INDUSTRIAL │            ─┤ RESOURCE MANAGEMENT │
          ─┤ INSTITUTIONAL │         ─┤ UTILITIES │
                                     ─┤ SPECIAL PROJECTS │
```

It should be pointed out that in a field as vast as the one covered here there are bound to be some overlaps. The powerhouse in a hydroelectric project, for example, is technically a building, but its nature is such that it would be classified under heavy construction rather than under industrial building construction. Conversely, a manufacturing plant could have a railroad siding, which would certainly not be considered as heavy construction under transportation. Thus a certain amount of discretion must be applied to what could be exceptions to the general intent of the classification.

Some of the items appearing in this classification are by present standards still somewhat exotic and may sound like science fiction. Yet these same items will be commonplace sometime in the not so distant future and by then will be completely taken for granted. For

1

BUILDING CONSTRUCTION

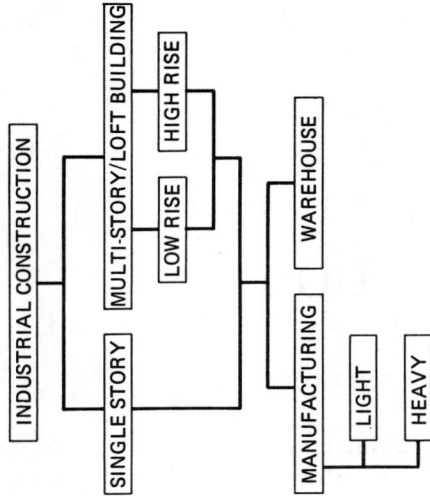

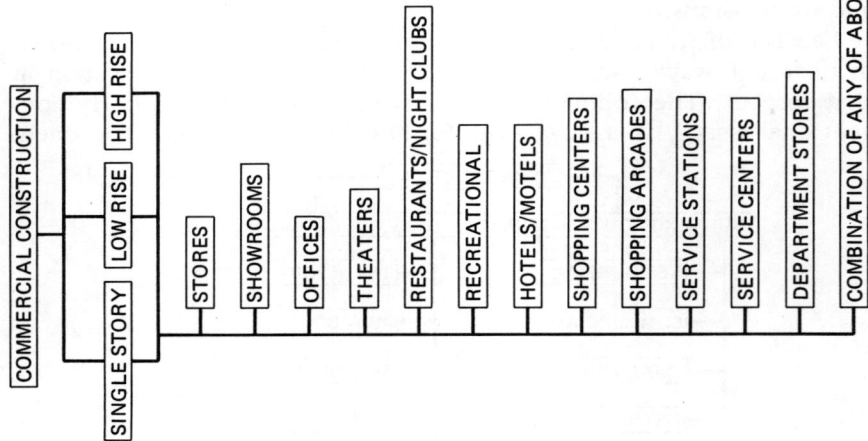

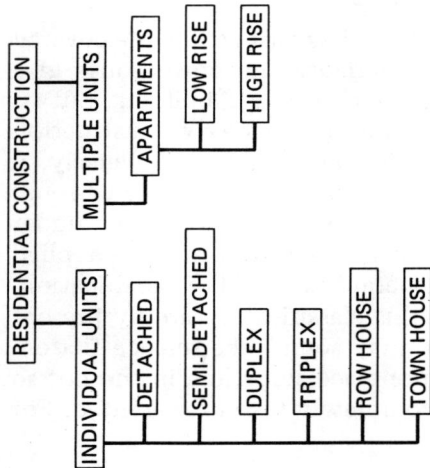

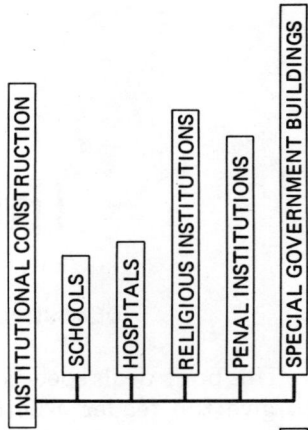

RESIDENTIAL CONSTRUCTION

- INDIVIDUAL UNITS
  - DETACHED
  - SEMI-DETACHED
  - DUPLEX
  - TRIPLEX
  - ROW HOUSE
  - TOWN HOUSE
- MULTIPLE UNITS
  - APARTMENTS
    - LOW RISE
    - HIGH RISE

COMMERCIAL CONSTRUCTION

- SINGLE STORY
- LOW RISE
- HIGH RISE
  - STORES
  - SHOWROOMS
  - OFFICES
  - THEATERS
  - RESTAURANTS/NIGHT CLUBS
  - RECREATIONAL
  - HOTELS/MOTELS
  - SHOPPING CENTERS
  - SHOPPING ARCADES
  - SERVICE STATIONS
  - SERVICE CENTERS
  - DEPARTMENT STORES
  - COMBINATION OF ANY OF ABOVE

INDUSTRIAL CONSTRUCTION

- SINGLE STORY
- MULTI-STORY/LOFT BUILDING
  - LOW RISE
  - HIGH RISE
  - MANUFACTURING
    - LIGHT
    - HEAVY
  - WAREHOUSE

INSTITUTIONAL CONSTRUCTION

- SCHOOLS
- HOSPITALS
- RELIGIOUS INSTITUTIONS
- PENAL INSTITUTIONS
- SPECIAL GOVERNMENT BUILDINGS

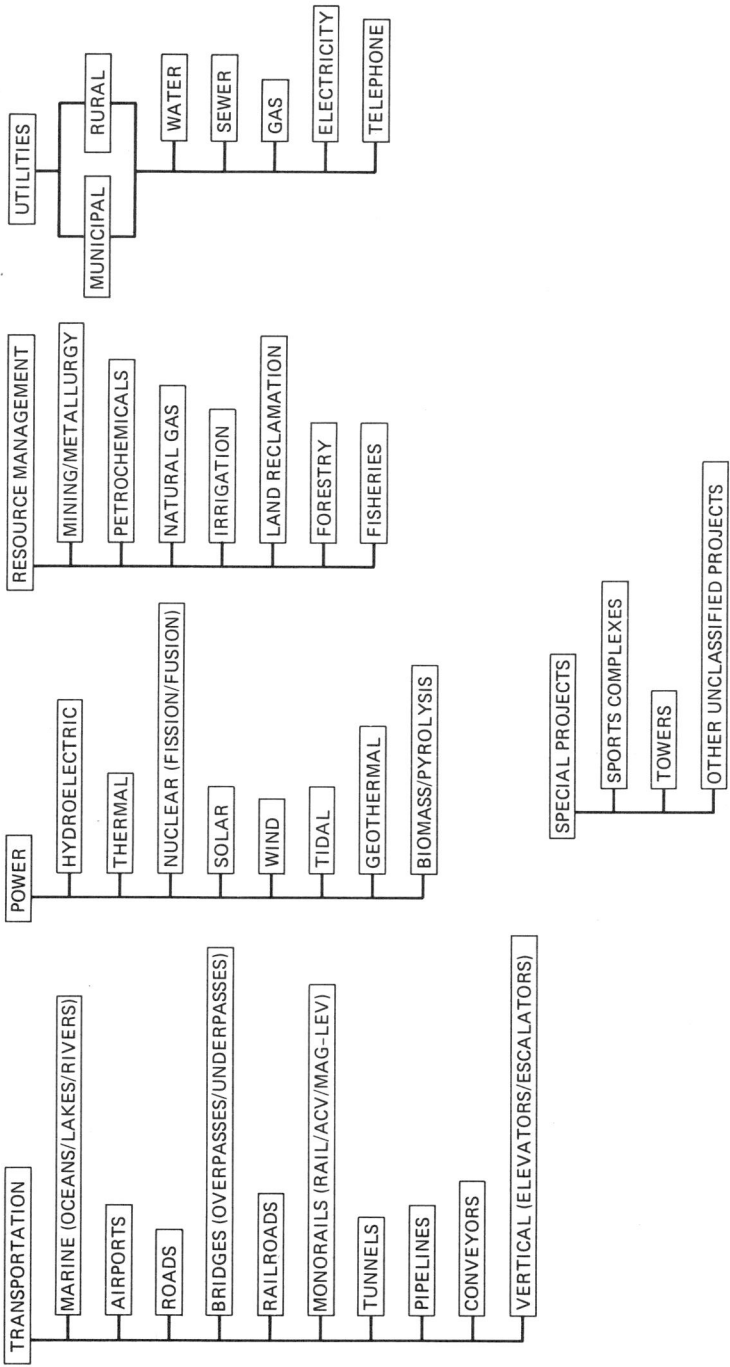

HEAVY CONSTRUCTION

TRANSPORTATION
- MARINE (OCEANS/LAKES/RIVERS)
- AIRPORTS
- ROADS
- BRIDGES (OVERPASSES/UNDERPASSES)
- RAILROADS
- MONORAILS (RAIL/ACV/MAG-LEV)
- TUNNELS
- PIPELINES
- CONVEYORS
- VERTICAL (ELEVATORS/ESCALATORS)

POWER
- HYDROELECTRIC
- THERMAL
- NUCLEAR (FISSION/FUSION)
- SOLAR
- WIND
- TIDAL
- GEOTHERMAL
- BIOMASS/PYROLYSIS

RESOURCE MANAGEMENT
- MINING/METALLURGY
- PETROCHEMICALS
- NATURAL GAS
- IRRIGATION
- LAND RECLAMATION
- FORESTRY
- FISHERIES

UTILITIES
- MUNICIPAL
- RURAL
  - WATER
  - SEWER
  - GAS
  - ELECTRICITY
  - TELEPHONE

SPECIAL PROJECTS
- SPORTS COMPLEXES
- TOWERS
- OTHER UNCLASSIFIED PROJECTS

the sake of the technically less informed reader the following explanations may be helpful.

Monorails, which to many people are still somewhat futuristic figments of the imagination, apart from the conventional rail-hung or rail-riding types, have alternative versions: ACV or air-cushion vehicles, in which the vehicles ride on a cushion of air but are guided on a central guide, or, alternately in a trough-like track; Mag-Lev (magnetic levitation) vehicles which are supported off the track and propelled by the action of a high-energy magnetic field.

Geothermal power takes advantage of hot springs or hot gases emanating from the earth.

Biomass and *pyrolysis* are based on utilizing the energy contained in organic waste and the conversion of substances chemically through the action of heat.

FUTURE TRENDS IN CONSTRUCTION

Apart from changes for strictly esthetic reasons, design will become influenced more intensely by such considerations as economics, ecology, energy, and other facets of life, some of which are on the way to reaching crisis status. This will tend to lead to many innovations and tax the designers' ingenuity. With the gradual realization of the limitations of natural resources and the global need for conservation it will become necessary to maximize efficiency in all areas.

It is interesting to note that the close approach and interrelationship between construction and manufacturing is gradually eroding what used to be previously a well-demarcated area. High costs, problems with organized labor, and construction in northern climates under severe weather and climate conditions have given special impetus to prefabricating as many building components as feasible. Thus actual field erection is reduced appreciably, permitting faster completion and increasing efficiency and profitability of the project.

The future has many surprises in store for us. We are already thinking of possibilities which are still only a gleam in the experts' eyes.

Excavation will be performed for large projects by applying nuclear methods with special radiation-proof equipment and methods.

Prefabrication will be much more advanced and a universal standarized code based on using metric modular units will be applied on a global basis. Not only will prefabrication permit assembly of building components manufactured in different locations at distant sites, but it will also be engineered so that various appliances and equipment, equally standardized, can be used anywhere, built-in if so desired, and with no worry about not being able to find spare parts or repair facilities locally in case of mechanical failure.

Many daily functions will become automated and computer-controlled, especially in residential construction. Lighting, heating, kitchen services, and many other functions will be timed to suit personal requirements and the complete integrated program will be capable of being triggered or adjusted by remote control commands given by telephone. General news, private messages, pictures, and even games will be projected on special wall and ceiling panels by means of built-in projection equipment. Many other functions in business and industry will be treated similarly. Miniaturization of formerly bulky items, thanks to modern electronic know-how, including microcircuitry and solid-state technology, will not only reduce size and space requirements of many building components control elements, but in view of the reduced power requirements will also help to conserve energy. All this will require special facilities and installations and will add greatly to the complexities of both design and construction.

Our changing energy needs and sources will affect building design greatly. Solar power will at first tend to be developed on an individual basis, with various houses or buildings having their own installations. This may be followed by complete solar power stations combined with underground hot water storage reservoirs and other ancillary facilities. Wind power will be utilized more in wind power stations. These, however, will only find efficient use in regions having sufficient wind currents, such as coastal areas and some mountainous regions. Coal, a long-time pariah among energy sources, will find new prominence, since the scrubbing and purification required to make it ecologically acceptable, will now be economically justified in comparison with the escalating costs of petrochemical and nuclear energy. Nuclear power stations may become commonplace for general power generation, but a lot of equipment will be powered by mini-nuclear power units, all mounted in accident-proof sealed units, which will outlast the equipment and be capable of being reused again and again.

Construction methods will change in all areas and manual labor will tend to disappear with more and more emphasis on mechanized operations. Thus there will be a demand for more specialists, primarily those having technical backgrounds.

Research and experimentation by inventive minds, both in the field and in the laboratory, will lead us into an era of progress in technology which will open new vistas and enable us to use our vast technical knowledge for the good of human beings throughout the world.

Building economics is a complex subject in which a number of considerations influence the ultimate design and construction criteria.

The building may be a speculative venture intended for resale after construction, or it may be kept as a revenue producer. In the former case, the tendency might be to save on capital investment and build to minimum specifications thereby reducing quality to a low standard. In the latter case, a long-term view must be adopted and capital investment must be balanced against maintenance and repair costs. At the same time, tax loss considerations, inflation trends, and comparisons of alternate investments for the capital sunk into the project must be investigated, resulting in a multidimensional pattern which may require a computer and possibly also a large crystal ball. When a policy decision has been reached, it must be applied to the design as well.

When one is considering the economy of a building in terms of design and construction a distinction should be made between the economy or feasibility of the overall project and that of the components of the building.

The overall economy is usually covered in initial feasibility studies and will govern the project to the point of go-ahead. At this stage the decision on the quality of the building has already been made within certain outlines and limits, but the economy of specific building components will usually be considered during the design and within limits during actual construction. Thus individual items must be judged and chosen based on their life cycle cost factor for long-term investments, i.e., the comparison of the relationship of the capital cost against the combined cost of maintenance and repairs, including replacements, must be correlated to the building as a whole and the resulting profile must be translated into equivalent quality building component design. At the same time, basic cost items must be balanced against the additional costs of providing visual esthetics, building prestige features, and somewhat more abstract aspects, with the end product fitting neatly into the projected budget range which will then become the starting point for further development.

Generally speaking, there are few projects going ahead in which the economy does not matter and in which "money is no object!" Yet regardless of the quality projected for the building there are at least two components in a building which receive more wear and tear than any others. They are (1) the doors and their hardware and (2) the wearing surfaces of the floors.

Since either of these represents a comparatively small percentage

6

of building costs (combined they amount probably to less than 10%), it is advisable and recommended to choose the best quality for either one; on a long-term basis false economy would be just what it implies. In this respect, it should be mentioned that since some buildings have as many as 20 different types of doors, economy should be exercised in the choice of doors with less use, but their hardware should generally be of the best quality.

The next item receiving high wear and tear is the mechanical installation, especially the air conditioning and heating systems, the more so when they are combined. Since, however, they may account for up to 30% of building costs, economy becomes a very definite factor, but it must be balanced with other considerations of a technical nature, such as flexibility, operation, maintenance characteristics, and so forth.

The economy of the structure depends primarily on the technical considerations which will govern the choice. The size and purpose of the building combined with subsoil conditions will probably dictate the type of foundation to be used. Floor loading, free spans, fireproofing regulations, among other considerations, will help to determine the type of structure chosen.

The economy of many other building components will be a function of the design regarding visual esthetics. Yet even here it is possible to encounter economic aspects. Let us take a high-rise office building with rounded corners as an example. The corners can be made by straight segments (Figure 1-1) following the curve, but if curved glass and sections are used to describe quarter circles, it would

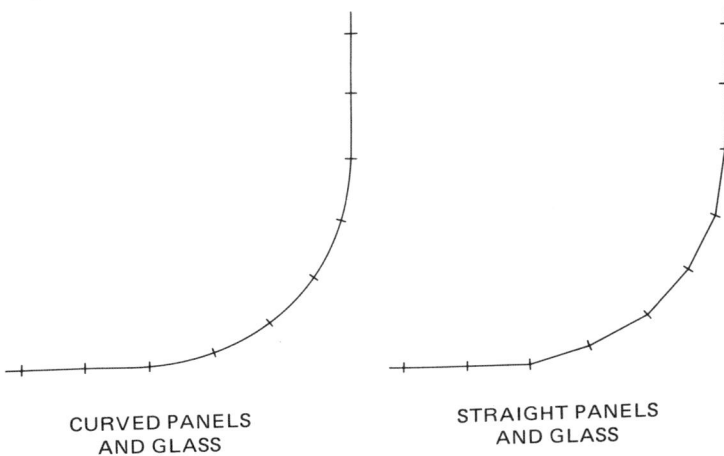

CURVED PANELS AND GLASS

STRAIGHT PANELS AND GLASS

FIGURE 1-1. Curtain wall-round corners. (a) Curved panels and glass are expensive. (b) Straight panels and glass are economical.

add an excessive cost to the windows and curtain wall. Similarly, the choice of the section and finish of the curtain wall sections could cover a substantial cost range. In either case, the overall effect from a visual standpoint would not change noticeably, especially when the building is viewed from some distance.

Thus economy must be practiced with discretion and prudence.

VALUE ANALYSIS IN CONSTRUCTION

Until comparatively recently many sectors of the manufacturing industry subscribed to the philosophy of planned obsolescence. Certain items were made to last for an intentionally limited time only because the demand for replacements would keep that particular industry busy in the future. As it was gradually realized that our resources are finite and with the resulting consciousness of the importance of conservation and recycling of resources, including materials as well as energy, an industrial technique known as *value analysis*, in close relationship with cost engineering, started to take a critical look at design and fabrication of many items. Although the construction industry cannot be accused of practicing planned obsolescence, it nevertheless has looked at most items only in terms of their edge relative to the competition, an attitude probably engendered by the competitive bidding systems of the past. With more freedom in design and changes in tender practices (procedures in calling bids for buildings), as well as the increased emphasis on long-term performance, the industry, starting with design, is looking more closely and critically at various materials as well as methods and judging them on terms other than only initial or basic cost. As the term implies, the choice goes to the item supplying the best value for the cost. In building construction this can be illustrated as follows: a building component or system is chosen. It is analyzed as to its function and purpose. Alternative materials, methods, and elements constituting the component or system are combined and costed for comparisons. The advantages and disadvantages of each are analyzed in turn and given values. Depending on the overall objective, the alternative showing the best value which could be on a cost/time ratio is selected and incorporated in the project. This is also known as *life cycle cost analysis*, a method which in comparing initial against continued cost factors or capital-investment and maintenance-replacement costs is intended to come up with the best alternative from a multiple choice on a long-term basis. In view of the complex nature of some analyses, these studies are usually computerized. The printouts may show that a certain amount of additional money spent over the apparent minimum necessary, in the long run, may result in considerable savings.

LOCATION OF BUILDING ON SITE

Few buildings can be located on their future sites without some restriction. In downtown or other built-up areas the lot will probably determine the maximum size, and building regulations will govern height, number of floors, setbacks, projections, and other physical features. Only when a building is on land which is much larger than the area to be covered by the building will it be possible to have more of a free hand.

Location of the building on the site may have visual aspects, but there are also technical considerations that must be taken into account during the design stage, primarily those dealing with climatic factors (Figure 1-2).

A southern exposure will mean high temperatures at noon during summer or, in practical terms, a higher air conditioning load plus possibly more glare.

A northern exposure will mean that this side will be colder and consequently represent a higher heating load which in turn could be offset by a corresponding decline in air conditioning load.

A western exposure will have a low sun in the afternoon and thus a somewhat higher air conditioning load combined with substantial glare.

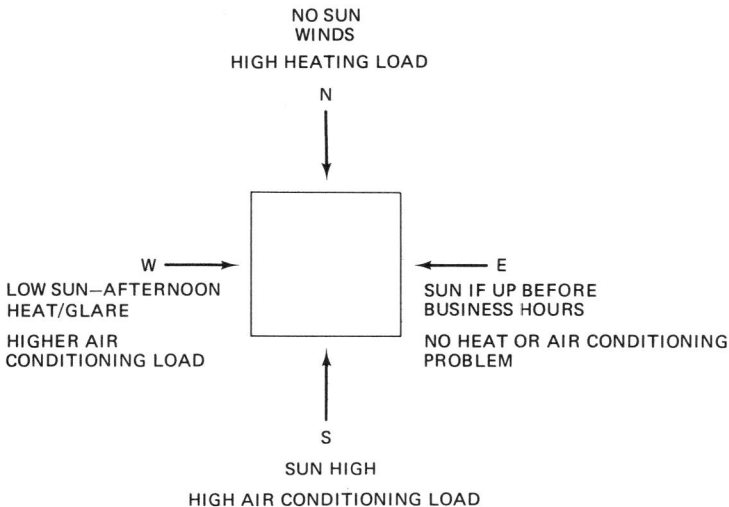

NO SUN
WINDS
HIGH HEATING LOAD

N

W →

LOW SUN—AFTERNOON
HEAT/GLARE

HIGHER AIR
CONDITIONING LOAD

← E

SUN IF UP BEFORE
BUSINESS HOURS

NO HEAT OR AIR CONDITIONING
PROBLEM

S

SUN HIGH
HIGH AIR CONDITIONING LOAD

FIGURE 1-2. Location of building on site.

9

An eastern exposure is of comparatively little significance on the North American continent because the sun is already fairly high by the time most daily activities or business life start and by about mid-morning it enters the southern quadrant.

There are a number of remedies to counteract the action of the sun. The most effective solution can be a tinted solar glass which may be quite expensive. It would probably be used throughout the building rather than on some sides only. There are some plastic films on the market which can be applied to the glass to cut heat and glare, but they are easily scratched, difficult to repair neatly, and at times may ripple because of the air bubbles expanding between them and the heated glass surface. Another alternative is to provide drapes or blinds, but this means reliance on people to close the drapes and blinds during intense sunshine periods and over weekends.

Lighting is usually not affected by these considerations since daylight is generally not too much of a factor in most commercial or industrial buildings where artificial lights are on for most of the day.

Another consideration is the location of the building with relation to parking and traffic flow on the site. This may influence location of the entrances.

Shopping centers have special aspects which will govern building location and orientation on the site. Although they may have huge tracts of land, most of which may become parking areas, there are technical considerations which have no relation to the building. Some of the larger regional-sized shopping centers may be built in stages, with department stores being built first and the retail store area following afterward. In the hierarchy of shopping centers the department stores are at the top of the list and may insist on maximum exposure, i.e., they may choose the prime locations, in which case the center may be designed to suit them. Consequently, this will govern locating the center on the site. Entrance locations will then be coordinated within the overall plan and in the case of a supermarket a car order pick-up will also be integrated into the master plan. Since most centers today are enclosed malls, the problems which used to plague the former open strip centers, such as a low sun in the afternoon on a western exposure, are no longer a factor.

In housing, considerations vary somewhat from other types of buildings. Living areas should be oriented by considering location relative to the sun and possibly also to local wind conditions. Orientation resulting in excessive heat loads should be avoided. This can be helped, too, by use of large overhangs and strategic planting of shade trees. In colder climates the glass area should be reduced on the wind side and possibly additional insulation may be required.

Placing a house in the center of the lot can be wasteful since often parts of the lot end up as nothing more than a passage and cannot be used for anything else. Keeping the house close to the street will provide more (Figure 1-3) privacy in the backyard and garden. Privacy is also achieved by keeping living areas at the rear and placing service areas in front. Stressing the appearance of the house from the street may result in disadvantages in interior space utilization which must always consider the three main areas: living, sleeping, and working.

ECOLOGY

Until a few years ago ecological aspects in relation to construction were virtually nonexistent. Some plants discharged polluted and/or polluting effluent into rivers and lakes, others spewed toxic and/or corrosive fumes into the air, still others dumped large quantities of noisome waste materials, but unless a specific complaint came up nothing much was done about it. Needless to say, this situation has changed and thanks to public awareness and pressure on government, action is gradually being taken on two related fronts: (1) reducing and eliminating harmful actions and materials and (2) recycling waste materials into active use. For construction, this must start right at the design stage in which many remedial actions can be taken.

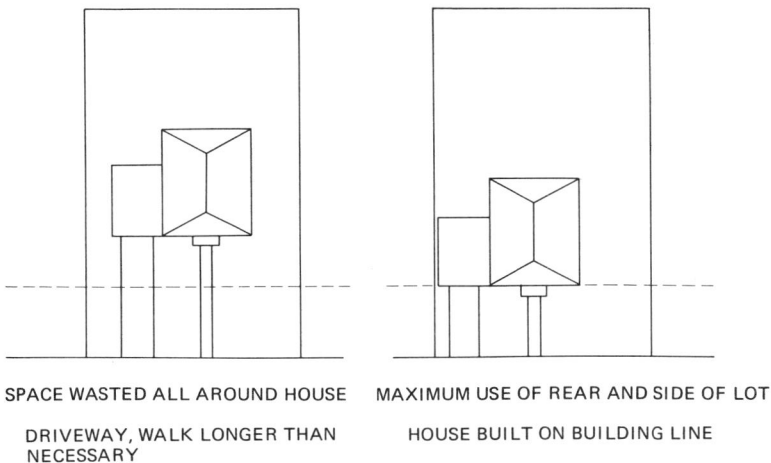

SPACE WASTED ALL AROUND HOUSE   MAXIMUM USE OF REAR AND SIDE OF LOT

DRIVEWAY, WALK LONGER THAN   HOUSE BUILT ON BUILDING LINE
NECESSARY

FIGURE 1-3. Locating house on lot.

Chimneys and exhausts can be equipped with scrubbers and/or electronic precipitators, effluents can be treated and neutralized before being discharged into the sewer system or elsewhere, and waste materials can be reclaimed and reprocessed.

There are, however, other aspects but they are not as well known or recognized. Some features of projects can create special localized problems, e.g., the huge parking lots of the regional-sized shopping centers act as large water catchment areas with the possibility of creating occasional overloads on the local sewer system. Large concentrations of exhaust gases accumulate at peak occupancy. The huge expanse of asphalt absorbs heat in summer and can affect localized climatic conditions.

Developers create additional problems by trying to develop land to its maximum use and by designing housing projects with excessive population density in order to increase their potential revenue.

Although ecological remedial measures are required in many areas, the construction industry, because of its size and importance, must take a very active part in this most essential aspect of what might become a fight for human survival in the future.

ENERGY

The more technologically advanced an industrialized society becomes, the more energy-intensive it will be, for machines rather than human or animal power sources will perform the heavy work. Thus the sources of energy are of extreme importance and with existing resources being used up at an alarming rate on a global basis, energy conservation has become a matter of economic survival.

Obviously, the energy crisis is here to stay and it will affect construction as much as other aspects of our lives. Construction has a dual relationship with energy. Construction consumes as well as helps to create energy or energy-producing installations. Because construction is highly mechanized, it can be an important user — and often waster — of energy.

Building construction which uses relatively little equipment is not as much affected as the highly equipment-intensive heavy construction.

Except for equipment used in excavation and foundation work, energy-consuming equipment is relatively limited, including among other items hoists, heaters, concreting equipment, generators, power tools, and so on.

As for the building itself, the designers must take care of energy

considerations. Since the energy crisis runs in parallel with corresponding cost increases, it is not only necessary to find new sources of
energy but also to conserve existing ones in the most economical
manner possible. Thus it is up to the ingenuity of the consultants to
convert liabilities into assets. Good examples of such techniques are
the use of waste heat from lighting fixtures for heating purposes or in
exploiting the thermal potential of flue gases.

Although we have the technological knowledge in many areas,
economic considerations are still stymieing us in using alternatives to
our conventional sources of energy. These include, among others:

A. Nuclear energy

B. Wind power

C. Solar energy

D. Tidal power

E. Geothermal sources

F. Biomass and so forth

Although the above are well within our capacity, apart from physical limitations of some of them, the biggest obstacles are of a political rather than economical or technical nature.

This book is neither trying to moralize nor is it taking sides for or
against certain controversial energy sources. It is, however, appropriate at this point to state a few practical facts. Because many people
believed and still do believe that there are unlimited energy sources
on hand, the choice of which energy source to use has almost always
been motivated first and foremost by economic considerations. Even
after the oil crisis brought the limitations of existing reserves of
natural energy resources into the limelight, it was mostly political
pressure that governed energy policies. Thus from a practical point of
view and assessing the problem on a long-term basis we realize that
we are faced with alternatives for which there is only a question of
time — and comparatively little time on a human scale — before even
the alternatives gradually disappear and we are faced with no choice
in either energy source or economic consideration.

Experts have pointed out that among natural resources oil reserves
seem to be good for another 50 years or so (on an optimistic scale)
but that coal reserves, and coal has been a pariah among energy
sources for some time now, would last us for several centuries. This
plus its relative position on the revised economic scale of energy costs
have given it a new prominence, which may disappear again if one of

the aforementioned alternatives comes into grace, either because it reaches a contemporary economic viability or because of sheer necessity.

VANDALISM

Vandalism in buildings has become a serious matter. The sad fact is that money has to be spent on unwanted items which may even result in spoiling visual effects. Nevertheless, vandalism must be considered in design and provided for. Although this applies perhaps more to public buildings than to others, all buildings, whether factories or private residences, could be affected.

The best precaution is to use a common-sense anti-destructive design which includes some of the following:

A. Public areas should avoid using vulnerable or soft materials. Fabrics can be ripped, plaster or plasterboard punched through, and projecting metal bent, broken, or ripped off. If these materials are used, they can sometimes be protected by sheets of acrylic or other plastics. Solid masonry, however, is preferable (Figure 1-4).

B. Decorations or artwork should not be delicate or they should have protective enclosures or fencing around them.

C. Sprinkler heads within reach should have guards of heavy wire mesh.

D. Exterior safety lights should have wire guards to protect them against target practice by slingshot-wielding young Davids who also see potential Goliaths in globular glass fixtures.

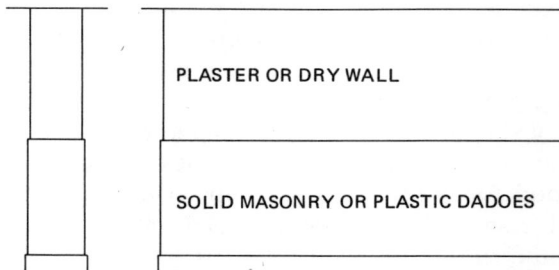

PLASTER OR DRY WALL

SOLID MASONRY OR PLASTIC DADOES

FIGURE 1-4. Protection of walls and columns.

E. Furniture should be heavy and bolted to the floor, and, if possible, have scratchproof finishes. Ash trays should be ballasted to make them more difficult to move or overturn.

F. Planters should not have delicate plants and flowers in them because they have a tendency to get decapitated or stolen.

G. Exterior water taps should either have removable handles or be in covered boxes that can be locked.

H. Show windows should be divided into smaller sections by vertical and horizontal mullions in order to reduce the cost of replacing broken panes of glass (Figure 1-5).

I. Garbage areas, a favorite playground for vandals, should also have special provisions against marauding animals. Containers should be well sealed, difficult to handle except by garbage removal equipment, and locked, if possible. Locking would require special arrangements with the garbage removal company as well as with the tenants using the facilities.

J. Public washrooms should have sinks without drain plugs, spring-loaded taps, and preferably electric hand dryers instead of paper towels, although dryers have been known to have been ripped off walls.

Although design can help to prevent vandalism, there must also be an efficient security service that makes regular inspections.

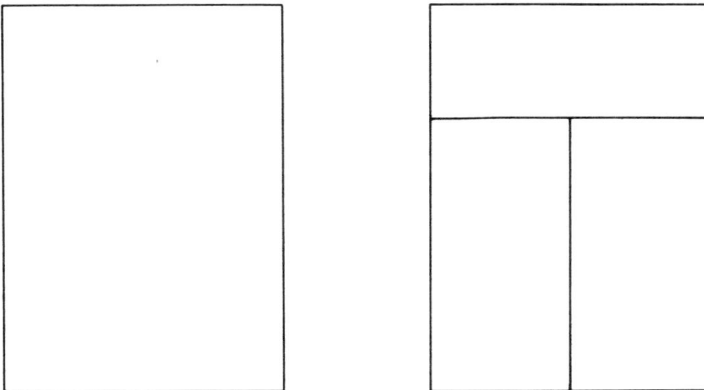

FIGURE 1-5. Show window: Reducing cost and probability of damage. (a) Higher initial cost; in case of breakage, higher replacement cost. (b) lower initial cost; in case of breakage, lower replacement cost because probability of damage is one-third.

SECURITY

It has become increasingly important to provide security for buildings regardless of size. This aspect has to consider the following:

A. Burglary and theft
B. Fire
C. Vandalism and sabotage
D. Crowd control

Burglary and Theft

Crimes as well as prevention are becoming increasingly sophisticated and the net effect manifests itself primarily in added building costs.

In private houses the usual countermeasures are to install burglar bars and a burglar alarm activated by forced entry. More expensive homes may have a direct line to a security force or to the local police indicating attempted break-ins.

Apartment buildings keep lobbies locked or have doormen around the clock to ensure that only authorized persons enter the building. Visitors to tenants must have identification for the same reasons. Garage doors are usually raised by key-activated operators or must be unlocked by hand to be opened.

Some motels and hotels use a fairly recent system which uses special coded cards instead of keys to open rooms or call elevators. Insertion of the card into a slot opens the door or activates the elevator. Since the combination or program for these cards is variable and can be changed easily, it is comparatively simple to maintain excellent security. The possibility of keys being duplicated or falling into the hands of unauthorized people is eliminated. Since stairs usually open only outward and cannot be opened from the outside on the ground floor, it is not possible for people to reach the private corridors leading to the rooms via the stairs.

In larger buildings such as office buildings visitors cannot be screened. These buildings may, however, in addition to security guards, have closed TV systems which monitor various areas of the buildings. These systems are often linked to private security services or the police for special emergencies.

Shopping centers or individual stores have special systems. Alarms connected to sensors on show windows (Figure 1-6) and storefronts are designed to be activated if the glass is broken or tampered with.

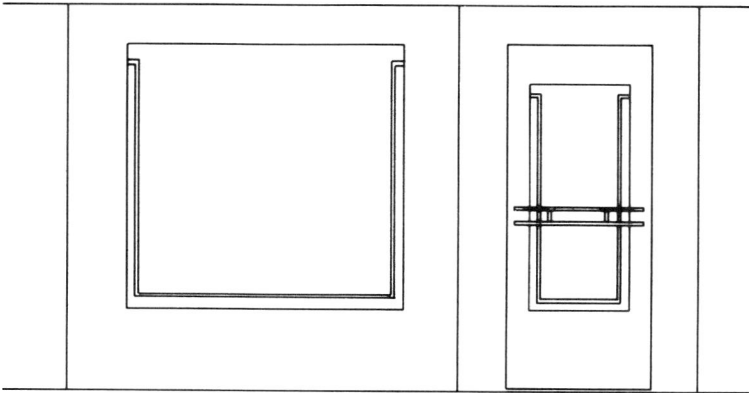

FIGURE 1-6. Burglar protection system on show window and door. A metallic strip completes the circuit. If the circuit is broken, an alarm is triggered.

The stores may have special magnetic anti-theft systems. The merchandise is tagged with special magnetized tags that activate a buzzer or bell when the merchandise is removed from the premises without being deactivated by the sales personnel. This system is usually installed separately and cannot be considered as part of the building.

Fire

Smoke or fire alarms are usually tied in with the fire protection systems and in the case of major buildings may be connected to the local fire department.

Private houses may have individual smoke or fire alarms which make a loud noise in case of fire.

Vandalism and Sabotage

This subject is covered in other chapters in this book. Senseless as they may be, vandalism and sabotage are taking on larger proportions all the time. Sabotage may enter the picture in special cases, particularly in buildings that house government agencies or representatives or in other buildings that may be the target of individuals or groups having grievances. Essentially, the only answer is greater vigilance on the part of the security staff because in most instances vandalism and sabotage affect public buildings which cannot be restricted effectively. A closed TV system that provides extensive coverage and that is supplemented by a PA system which could possibly be used in a deterrent capacity may be of help.

Crowd Control

There are two ways to control crowds: either by using security guards, especially at mass gatherings or special events, or, for normal conditions, by appropriate design measures. In shopping centers, for example, crowds can be directed by strategically locating kiosks or by using decorative features such as planters, seating areas, and so forth in the mall. Installing temporary barriers can always be done on a limited basis, but this has nothing to do with building design. The important point to remember is that any controlled traffic flow must not become a bottleneck in case of a real emergency because a bottleneck could create disastrous results.

HANDICAPPED PERSONS

In the past building designers have neglected to consider handicapped persons, specifically persons in wheelchairs, but today these people are getting more recognition. Although it does mean a certain additional expense, this is negligible in terms of total building costs. It is more a question of providing for circulation and use of facilities for persons confined to wheelchairs.

Following are some of the provisions that should be considered:

A. Sidewalk ramps for exterior curbs and sidewalks at parking lot. The ramp should preferably not be steeper than with a slope of 1:4, have anti-skid finish, and be painted a bright color for extra visibility. No-parking signs should be installed on either side of the ramps.

B. Ramps may also be required in low parking garages which have no elevators (Figure 1-7).

C. Elevators should be provided in buildings having more than one story, such as shopping centers or department stores. Elevator doors should be wide enough and be timed to stay open long enough to permit entry of the wheelchair without its getting caught. Control and call buttons should be low enough so that the person in the wheelchair can at least reach the ground floor button.

D. In washrooms special cubicles should be allocated to wheelchair

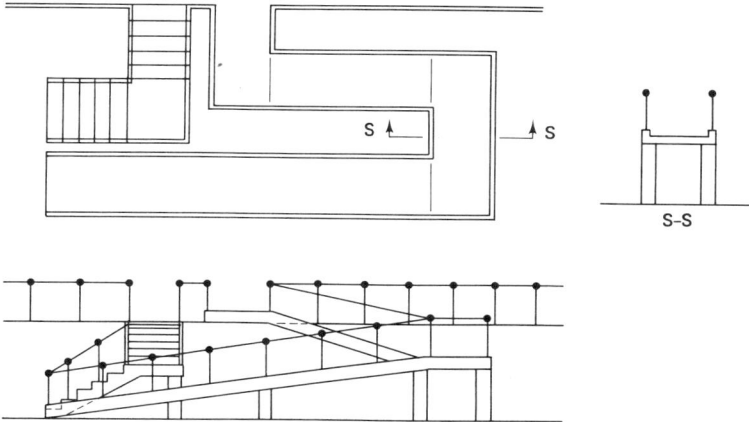

FIGURE 1-7. Ramp for wheelchairs.

passengers and should be marked accordingly. They require extra wide doors and enough room inside for the wheelchair to maneuver.

E. Passages should be wide enough to accommodate wheelchairs and there should be an area available for turnaround.

F. Entrance doors should have at least one door wide enough and easy enough to open by the wheelchair passenger.

G. Light switches should be within reach.

H. Public telephones should be provided at wheelchair height.

I. In private residences provisions are sometimes made for a small elevator. Another installation could be a stair elevator which consists of a traveling seat which follows a wall-mounted rail up the stairs. The latter, however, is not really suitable for a fully handicapped person.

CONSTRUCTION FAILURES

It is interesting to note that even thousands of years ago there existed building codes which included provisions in case of building failures, with penalties on the basis of an "eye for an eye" relationship. Chances are that the unfortunate builder to whom this happened, and who in those days was responsible for design as well as for construction, blamed evil spirits or the wrath of the gods for his misfortune in the failure of his work.

Building failures often occur during construction, but many may happen at a much later date, either because of latent defects, normal aging and deterioration, or extraneous factors such as earthquakes, lightning, storms, earth slides, etc.

The failure is generally due either to design, affecting the architect and consulting engineers, especially the structural engineer, or it happens during construction, in this case affecting the contractor or builder if it is due to faulty methods or the nature and quality of the building components or materials.

There are various reasons for failures, but the following cannot be condoned under any circumstances:

- Carelessness, negligence, or plain ignorance
- Incompetence of personnel in charge
- Taking chances or cutting corners in order to save on cost, time, or quality
- Lack of professional supervision

Other reasons include unanticipated causes beyond the control of either designers or builders. These are discussed later in this section.

Any building failure, unless it is a clear-cut case, will tend to involve and implicate a number of parties including:

- Owner
- Architect
- Consulting engineers
- Contractor or builder
- Supervisor in charge
- Inspector
- Building permits department

Often the responsibility is far from clear-cut and design and construction problems may overlap. Any such case ending up in court may take years to resolve and is likely to go through all avenues of appeal.

The majority of building failures are of a structural nature. Following is a list of the more common causes as related to different materials and construction methods:

Foundations:
- Undersized footings resulting in settlement of subsoil
- Uneven loading of footings

- Ground water erosion causing subsidence
- Piles generating insufficient bearing or friction value
- Piles being overloaded
- Tubeless piles failing because of "necking" (Figure 1–10)
- Wood pilings being attacked by fungi, vermin, etc.
- Wood piles rotting on intermittent exposure to air and moisture (Figure 1–11)
- Foundations sitting on geological slippage plane
- Excavation removing support of adjoining structure without adequate underpinning or shoring (Figures 1–8 and 1–9)
- Vibrations set up from blasting operations or, in special cases, from heavy passing traffic
- Draining of water at distant site affecting and lowering local ground water table
- Unstable foundations resting on sandy soil on hillsite.

Formwork:
- High forms inadequately braced or shored
- Formwork overloaded
- Formwork unevenly loaded during concreting (Figure 1–12)

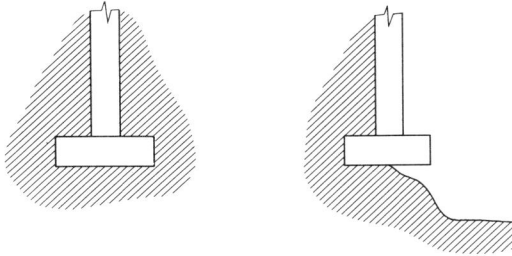

FIGURE 1–8. *Effect of unconfined excavation on adjoining footing. Subsidence under footing can cause failure.*

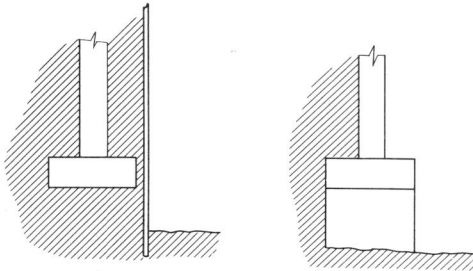

FIGURE 1–9. *Protection of adjoining footing by sheet piling or underpinning.*

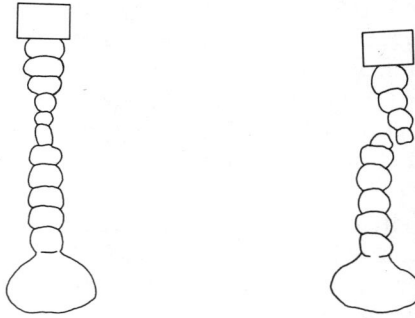

FIGURE 1-10.   *Expanded base tubeless pile failing by "necking."*

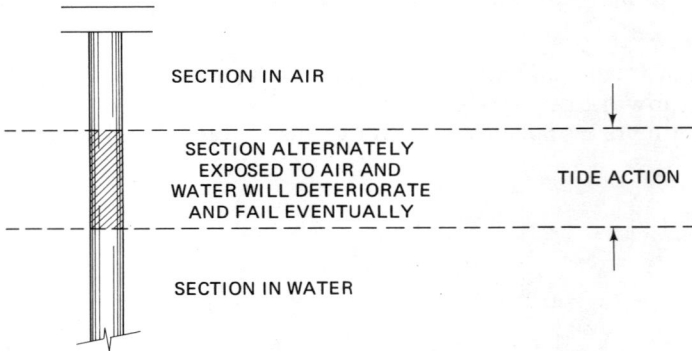

SECTION IN AIR

SECTION ALTERNATELY
EXPOSED TO AIR AND
WATER WILL DETERIORATE
AND FAIL EVENTUALLY

TIDE ACTION

SECTION IN WATER

FIGURE 1-11.   *Wood pile exposed to air and water.*

- Concrete poured too quickly creating impact stresses
- Vibrating concrete creating a critical frequency in formwork members
- Improper design of forms or falsework
- Premature removal of forms
- Failure to reshore adequately
- Inadequate or unfirm base for shores and jacks
- Shores and jacks off vertical not developing full capacity
- Failure to allow for higher erection stresses during construction

Concrete:
- Understrength concrete
- Frozen concrete

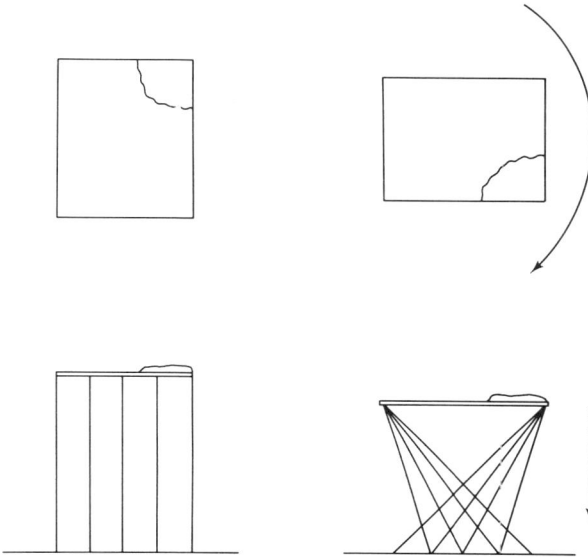

FIGURE 1-12. Failure of slab form. Deck form is on very high supports, which are not adequately cross-braced. With concrete being placed in one corner, the uneven loading of the deck causes the deck to rotate and as a result, the supports and deck collapse along a rotating pattern.

- Cold joints in structural members
- Overstressing building components with shoring or formwork for overlying structure
- Insufficient temporary bracing or shoring for precast members
- Failure of pretensioning wires
- Failure of posttensioning anchorages
- Failure of precast member connections or supports
- Eccentric loading during concreting
- Eccentric loading due to permanent or temporary materials — landscaping, snow loads, etc.
- Concrete spoiled by (accidental) addition of harmful substances
- Inadequate bracing of lift-slab columns

Structural Steel:
- Failure of members in shear, tension, compression, or torsion
- Failure of connections
- Insufficient bracing and guying during erection before final welding and bolting

- Inadequate design for excessive erection stresses
- Corrosion

Masonry:
- Inadequate foundations
- Mortar not developing bond
- Tipping due to wind loads

Timber:
- Failure of connectors or connections
- Timber attacked by sea water, soil, vermin, dry rot, etc.
- Timber intermittently exposed to air and moisture
- Overstressed members
- Unsound quality of wood used

Roof:
- Faulty design
- Faulty workmanship
- Inadequate provisions against movements in structure
- Installation or materials affected by weather conditions

Among unanticipated and sometimes unprecedented causes for failures are natural disasters such as earthquakes, landslides, hurricanes, tornadoes, rainstorms and floods, sudden large drops in temperature, and so forth.

At times failures may occur under strange circumstances. A stepped storage building failed because it was overloaded with a material whose angle of repose did not follow the slope of the walls. Because of the overfilling the bottom of the walls was literally lifted off the foundations.

Thin shells are especially vulnerable to uneven loading. The greatest danger is caused by snow loads which can vary depending on local wind conditions and under changes of temperature. In milder weather melting snow becomes extremely heavy and causes many unanticipated failures. Wind loads, too, can be uneven and create special hazards, especially if they induce critical frequency in tall structures. Critical frequency is also an important factor in earthquakes and must be considered accordingly during the design stage.

Certain failures, especially in steel structures or precast concrete frames, often have a domino effect. Failure of one member tends to induce other members to fail in a chain reaction, which is often due to the large negative stresses generated or the sudden and excessive impact stresses.

Temporary platforms and grandstands, which are seldom properly engineered, have often been the scene of failures caused by their exposure to moving uneven loads.

Older buildings often are in a state of equilibrium. When alterations are made in them this equilibrium may be upset and in extreme cases can result in failures of some building elements. Failures may also be caused by the fact that changes made in the past are not known or apparent and show up only during renovations, or worse, during the failure. In such cases, a latent defect can also cause a delayed failure by upsetting the latent equilibrium of the building.

It may perhaps sound pessimistic but chances are great that many more building failures will occur in the future. Although building codes are becoming more stringent (in fact, in many places even formwork must be designed by professional engineers) and supervision and inspection are improving, there are other factors such as our air pollution which will become a major villain. One of the main problems may be corrosion of structural steel and reinforcing steel, failure of connections of precast members, failure of prestressed members, and deterioration of concrete, mortar, and other vital building components. Only rigid periodic inspection and timely countermeasures will prevent these disasters which are there waiting to happen.

METRICATION

Metrication refers to conversion to the SI (Système International) system which is the universal system of using decimal interrelated units of measurement.

There is no question that it is only a matter of time until metrication is completely universal and all countries are standardized on the decimal system. On this continent Canada has already started but the United States is still holding out although this is likely to change in the foreseeable future. Most European countries have been metric all along, as has South America. Great Britain was perhaps the only exception, but it too is following the trend. In Asia and Africa the only countries which are not metric yet are remnants from the old colonial days and they too will eventually convert.

The advantages of the metric system are obvious. Apart from the simplified calculations enabled by the decimal system, the interrelation between various units makes conversions that much easier. The most difficult aspect of metrication once it is implemented will be to get used to it and to unlearn the imperial system. Visualizing units of the decimal system on a quantitative basis will take some getting used to and will take time and practice. Modular building systems will be both easier to design and build. The current reluctance to convert is purely economical because the industry objects to the costs of converting and/or retooling.

In order to ease the problems of metric conversion there are two possible alternatives of converting.

"Soft Conversion" utilizes existing standards until further notice, but divides them into decimal or metric equivalents, for example $4' - 0'' \times 8' - 0'' = 1219$ mm $\times$ 2483 mm.

"Hard conversion" which will change standards to metric sizes immediately, for example $4' - 0'' \times 8' - 0''$ will become 1200 mm $\times$ 2400 mm.

It is beyond the scope of this book to treat metrication in depth, but some of the more common conversion factors which may be encountered in building construction are given in the following table:

Metric Conversion Table

Read: ⟶ Multiply to Get	By By	to Get Divide ⟵ Read:
Distance:		
Inches	2.54	Centimeters (cm)
Feet	30.48	Centimeters (cm)
Feet	0.3048	Meters (m)
Yards	0.9144	Meters (m)
Miles	1.61	Kilometers (km)
Area:		
Square inches	6.452	Square centimeters (cm^2)
Square feet	929	Square centimeters (cm^2)
Square feet	0.0929	Square meters (m^2)
Square yards	0.836	Square meters (m^2)
Square miles	2.592	Square kilometers (km^2)
Volume:		
Cubic feet	0.0283	Cubic meters (m^3)
Gallons (U.S.)	3.7854	Liters (l)
Gallons (IMP)	4.546	Liters (l)
Mass:		
Ounces	28.35	Grams (g)
Pounds	0.4536	Kilograms (kg)
Pressure:		
Pounds/square inch	0.0705	Kilograms/square centimeters (kg/cm^2)
Pounds/square foot	4.883	Kilograms/square meter (kg/m^2)

Metric Conversion Table (Continued)

Read: ⟶ Multiply to Get	By By	to Get Divide ⟵ Read:
Density:		
Pounds/cubic foot	16.05	Kilograms/cubic meter (kg/m^3)
Speed:		
Miles/hour	1.61	Kilometers/hour (km/h)
Miles/hour	0.447	Meters/second (m/s)
Feet/second	0.3048	Meters/second (m/s)
Temperature:		
Fahrenheit ($^\circ$F) (F - 32)$\frac{5}{9}$ = C		Celsius ($^\circ$C) (Centigrade)

Note: Celsius does not belong to the SI System, but it is metric.

Metric System Conventions

Dates: Year — Month — Day — Hour — Minute — Second.
Example: 1979 — 01 — 06 — 16 — 24-08 would represent 24 minutes and 8 seconds after 4:00 P.M. on January 6, 1979. Format permits dating to exact second, although usually only day is required.
Numbers: Spaces are used after every three digits.
Example: 12 000 000 *not* 12,000,000
Comma is used to indicate decimal. (Period may be used in Anglophone countries.)
Example: 12,34 (12.34)
Zero is required for numbers smaller than one.
Example: 0,56 (0.56) *not* ,56 (.56)
Units: If number is spelled out, unit is spelled out.
Example: Four meters or 4 m *not* four m
Units are always in singular.
Example: 5 cm *not* 5 cms
Area: cm^2 *not* sq. cm.
Volume: cm^3 *not* cu. cm. or c.c.

2

Materials

As with other aspects of construction, materials tend to become more and more diversified. In addition, whereas formerly various building materials appeared on the job site as the basic elements of the final building, the trend is gradually more toward prefabricated, ready-made and often custom-built components, which can be relatively easily and quickly assembled and installed on the job site. Thus the tendency is to turn from site-built to factory-built items, which means greater emphasis on manufacturing.

Materials can be divided on a very general basis into structural and nonstructural materials.

Structural materials include those which are used to hold up and support the building; nonstructural materials can be considered as the balance and their purposes are mainly functional and esthetic.

Structural materials consist primarily of the following groups:

- Wood, as used in building framing and laminated structures
- Steel, as used for structural steel and reinforcing steel
- Cement and concrete, as used in concrete construction
- Masonry, as used in bearing walls and other supporting construction elements.

It should be noted that there are overlaps both within this division and within the so-called nonstructural materials.

Concrete blocks, although made of concrete, are generally considered masonry material.

Decorative precast concrete panels are not necessarily structural elements of the building.

Precast beams, columns, and slabs, however, are structural items.

Steel as used for some miscellaneous ironwork which is not part of the building structure may nevertheless perform structural functions. Metal stairs would be one example.

Thus the foregoing classification must be used with discretion and a certain latitude in definition must be understood.

When materials are being chosen for design purposes a number of aspects must be considered, including the purpose and function of the material in its ultimate use, the restrictions the material will encounter, the cost/economy ratio of alternate suitable materials as well as the advantages and disadvantages of either, the durability to be expected, ease of installation, maintenance required, aspects of repairs or replacement, and other factors which may be specific for different items.

The following sections expand to varying degrees on some of the more common construction materials and their use in different building components.

WOOD

Wood is the only structural material being used in its natural state, except for any preserving treatment which might be applied.

Since the grade and quality of wood depend on geographic and climatic variables, it is the least homogeneous of the structural materials and due allowance has to be made for this in structural design using wood.

Wood consists of a complex formation of cellulose arranged in fibers in elongated shapes held together by lignin. These fibers are longer in softwood, reaching about $\frac{1}{8}$ in. (3 mm), and about one-third of that in hardwood. Their length is of the order of 100 times their diameter.

Wood is subdivided arbitrarily into two main groups: hardwoods and softwoods. The difference actually does not relate to their hardness or softness but is based on a different concept. Hardwoods generally come from trees which are deciduous, whereas softwoods come from coniferous or evergreen trees. In addition, trees of specific types, occurring in different regions, show such pronounced differences that they have additional identification, often by color or of a geographical nature.

Wood used for structural purposes often is indigenous to the region of the construction project, mainly because it will be better adapted to the local climate and other conditions and it will be more economical.

In order to understand the various properties of wood it is appropriate to look at the anatomy of trees and the factors which cause the variations in types and grades of lumber.

A typical cross section of a tree shows the following: Figures 2-1 and 2-2. At the core of the tree is the pith, which consists of dense inactive cells. This is surrounded by the heartwood, which is fairly dark and dense. The next layer is the sapwood, which is lighter and not as dense and is the active part of the tree. The outermost layer is the bark, which consists of an inner and outer layer. The zone between the sapwood and the inner bark, the cambium, is the area where the tree actually grows. It is there that the annual growth rings are added. It should be noted that there is a marked difference in the new wood depending on whether it is springwood or summerwood. The springwood grows faster and has less density, it is lighter in color, and it has less strength than the summerwood, which grows slower, is denser, and appears darker in color.

A narrow ring indicates a short dry season; a wide ring indicates better growing conditions for this particular year. Generally, narrower rings and a larger proportion of summerwood will result in stronger wood. Since the thickness of the annual rings is also a function of the rainfall during that particular season, a cross section of the tree indicates the pattern of precipitation. It is possible for a tree to develop two rings in one season. The first ring could be caused by a drought that would stop the growth; the second ring could be caused by resumption of growth after the drought. But this is a rare occurrence.

All trees start with growing sapwood, which performs an important function. The fibers in the sapwood act as transportation

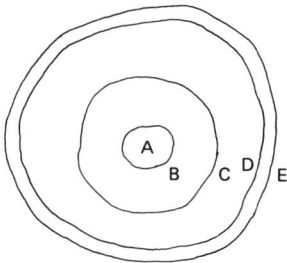

A: PITH
B: HEARTWOOD
C: SAPWOOD
D: CAMBIUM
E: BARK

FIGURE 2-1. Cross section of a tree.

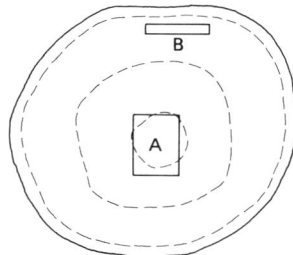

FIGURE 2-2. (A) Cutting beams or (B) boards.

channels for the plant food from the roots up through the trunk to the branches and leaves or needles of the tree. Thus the sapwood and the bark are the active parts of a tree.

Under adverse conditions the heartwood and pith become inactive and are the first to rot. This is seen at times in hollow trees that leave an apparently healthy exterior.

The tabulation at the end of this section lists some of the hardwoods and softwoods found on the North American continent, including various subdivisions of some of the main species. A number of exotic trees have been and are being transplanted on this continent with varying degrees of success.

Only a number of woods appearing in this list are used in construction; some are used for furniture, others have specialized uses such as for sports equipment, and many have no specific commercial or industrial use except for landscaping.

Wood is a hygroscopic material. Its cells absorb and give up moisture depending on humidity conditions, and in addition wood contains free moisture in the interstices. When continually immersed in water, wood will not deteriorate, but if it is exposed to air intermittently, such as timber pilings on a wharf, it will eventually rot and decay. When absorbing moisture it tends to swell and when drying out it shrinks, primarily across the grain. When the free moisture has evaporated or been given up, wood reaches the *fiber saturation point* and on further drying will start to shrink.

Woods vary greatly in density, ranging from 25 to 65 lb/cu ft (0.40 to 1.05 gm/cm³). Hardwoods generally are heavier than softwoods since they are denser.

The average moisture content of green wood varies greatly. Although the fiber saturation point of most species ranges between 20% and 30% of kiln-dry weight, the moisture content of green wood may exceed the weight of the wood itself by a sizable margin, especially in softwoods.

The strength of wood is greatly affected by knots, checks, shakes, splits, or other imperfections and by the amount of seasoning it has undergone. Very irregular grain formation, which is a function of its past growth pattern, also can have adverse effects.

Since wood has good mechanical properties, it is suitable for structural purposes, though on a much more limited scale than other structural materials. It is good in tension and compression parallel to the grain and in shear and bending across the grain. In addition, its hardness makes it suitable for other purposes such as flooring or furniture. It is an excellent insulator against heat and against electricity when dry. When it is wet, wood transmits heat faster and it conducts electricity. Its strength tends to increase with its dryness. Excessive sap content will tend to reduce its strength. In order to be usable in construction, lumber must be seasoned or dried. This is done either by air-drying, with lumber stacked in the open and left to dry on its own, or by kiln-drying, whereby it is dried under carefully controlled heat and humidity conditions in a kiln. In either case, the lumber is stacked in an open pattern to permit air circulation around each piece.

Air-dried lumber having a moisture content of about up to 15% is suitable for construction. Kiln-dried lumber, which obviously is more expensive, is used more for purposes which require a much lower moisture content, such as for flooring or furniture.

The National Lumber Manufacturers Association has established a number of definitions and standards which are generally used in the industry to grade lumber.

Yard lumber is generally used in wood frame construction. Shop lumber is similar but the pieces are shorter. Shop lumber is also used for manufacturing. Structural lumber is used for heavy beams and posts, which are usually cut from the heartwood of trees to utilize the strongest part of the tree. A distinction is made between rough and dressed lumber. Dressed lumber loses from $\frac{1}{4}$ in. to $\frac{1}{2}$ in. (6 mm to 12 mm) in machining. Any dimensions usually listed are nominal only and must be adjusted for both in design and in construction in the field. Additional distinctions are made between worked, matched, patterned lumber, shiplap, tongue and groove, and other machined shapes.

Boards are defined as being up to 2 in. (5 cm) thick; joists and planks are defined as being from 2 in. (5 cm) to 5 in. ($12\frac{1}{2}$ cm) thick and 4 in. (10 cm) or more wide; beams are defined as being 5 in. ($12\frac{1}{2}$ cm) or more thick and 8 in. (20 cm) or more wide, with a load applied on the narrow edge; and posts are defined as being 5 in. ($12\frac{1}{2}$ cm) or more in a roughly square cross section, stressed in compression.

Lumber can have natural defects such as checks, shakes, wanes, splits, knots, pitch and bark pockets, and so forth. All of these reduce the strength of the lumber and are considered in design within certain permissible limitations. In addition, wood can be attacked by certain diseases caused by fungi, such as rot, molds, decay, and so forth. Insects such as termites, carpenter ants, teredos, and other wood-boring pests destroy wood, but the major enemy of wood is and remains fire.

In order to protect and preserve wood, other than by external means such as painting, the following methods are primarily used:

- Chemical applications, such as Pentox, which are absorbed into the fibers
- Creosoting, which involves application of a coal-tar derivative distillate and is used for wood exposed to the elements or embedded in organic soil
- Osmose process, in which the preservative is applied to the wood in a vacuum chamber under carefully controlled pressure and temperature conditions (softwoods generally are easier to treat than hardwoods)

Lumber is also used in built-up laminated sections. These have the advantage that shorter pieces can be utilized, which may be wasted otherwise, but the pieces require more meticulous machining. They must be absolutely smooth and without the slightest warp in order to ensure perfect bonding of the individual layers. The glue has to be of extremely high quality and able to stand up to the variations of humidity as well as to the large range of temperature.

Plywood, which is finding continuously new applications in construction, consists of thin layers of veneer combined in a cross-banded manner, which means that alternate layers have the grain at right angles to each other. The veneer is often cut from trees by using a rotary cutting method which peels the tree trunk in the form of a spiral. In order to prevent the veneer from splitting, the wood must not be too dry and the logs from which the veneer is cut are usually kept in wet storage and only dried partially for cutting. Plywood is usually made in an odd number of layers. By having the grain run in the same direction on both faces helps to equalize stresses and prevents the sheets from warping. Nevertheless, they should be stored horizontally.

Depending on its ultimate use, plywood is graded in different ways including (A) combinations of good, sound, or solid, (B) one or two sides, which refer to the quality of the veneer and whether holes and cracks are patched, and (C) grades such as marine, sheathing, and so forth. Plywood weighs on the average around 3 lb/sq ft for a 1-in. thick sheet (approximately $14\frac{1}{2}$ kg/m^2 for $2\frac{1}{2}$ cm thickness) with thinner sheets roughly in proportion. Hardwood veneer plywoods are slightly heavier than those made from softwoods.

As an alternative to laminated sections in wood structures, such as trusses, these can be built up with regular pieces using special timber connectors. The most common ones consist of either split ring or toothed-ring connectors that are fitted into grooves cut into adjoining members by means of a special grooving tool. They are used in combination with bolts and steel plates. Each connector has a special

HARDWOODS AND SOFTWOODS

Hardwoods:

Ailanthus
Alder
Ash
   Biltmore white
   Black
   Blue
   Brown
   Green
   Oregon
   White
Aspen
Basswood
Beech
Birch
   Paper
   Sweet
   Yellow
Buckeye
Butternut
Cherry
Chestnut
Cottonwood
   Eastern
   Northern black
Elm
   American
   Rock
   Slippery
Ginkgo
Gum
   Black
   Red
   Tupelo
Hackberry
Hickory
   Bigleaf shagbark
   Mockernut
   Nutmeg
   Pecan
   Pignut
   Shagbark

Softwoods:

Balsa
Cedar
Alaska
   Aromatic red
   Eastern red
   Incense
   Northern white
   Port Orford
   Southern white
   Western red
Cypress
Douglas Fir
   Coast region
   Inland Empire region
   Rocky Mountain region
Eucalyptus
Fir
   Balsam
   Lowland white
   White
Hemlock
   Eastern
   Western
Juniper
Larch
Laurel
Pine
   Idaho white
   Loblolly
   Lodgepole
   Longleaf southern
   Northern white
   Norway
   Ponderosa
   Pinyon
   Shortleaf southern
   Sugar
   Western white
   Yellow
Redwood
Sequoia

HARDWOODS AND SOFTWOODS *(continued)*

Hardwoods: *Softwoods:*

Hickory (continued)
    True
    Water
Linden
Locust
    Common
    Honey
Magnolia
    Cucumber
    Evergreen
    Swamp
Mahogany
Maple
    Bigleaf
    Birdseye
    Black
    Drummond
    Hard
    Norway
    Red
    Silver
    Soft
    Sugar
Mulberry
Oak
    Black
    Burr
    Chestnut
    Laurel
    Pin
    Post
    Red
    Scarlet
    Southern red
    Swamp chestnut
    Swamp red
    Swamp white
    Water
    White
    Willow

Spruce
    Black
    Eastern
    Engelmann
    Norway
    Red
    Sitka
    White
Tamarack
Yew

Hardwoods (continued)

    Osage
    Plane
    Poplar
    Sycamore
    Tulip
    Umbrella
    Walnut
    Willow

design value, and joints having large stresses may require several connectors which are arranged in special patterns following established design procedures.

CONCRETE

Just as wood is the most widely used natural material, concrete is the most widely used manufactured material in building as well as in other types of construction. Concrete is also one of the oldest materials. Cement was used thousands of years ago. The use of concrete became more prevalent around the turn of the century in Europe and somewhat later on the North American continent.

Concrete is a mixture of substances which when mixed together with water combine in chemical action to create a hard, durable, and compressively strong material which is used primarily in building structures, but it is also used for architectural building components.

The basic ingredients of concrete are a mixture of cement, fine and coarse aggregate, and water. In addition, there may be admixtures for waterproofing, coloring, increasing durability, accelerating or retarding the setting, or other purposes.

Special cement may be used, such as high-early strength cement or colored cements.

In mixing concrete the mixing time must be kept within a certain range and the water content must conform to the amount specified in the design mix. Increasing the water content makes the concrete easier to handle and place, but it will also result in a reduction of concrete strength.

Aggregate may be used for special visual effect in exposed concrete work, which is achieved either by sandblasting or chemical treatment of forms. Setting the aggregate in the surface of the concrete before it has reached its initial set creates a mosaic effect.

Concrete becomes stronger as it ages, but the increase in strength levels off after a certain time. Because of this variable strength, regular concrete is referred to by its strength at age 28 days. Basic tests of concrete are the slump test taken at the time of placing the concrete and compressive strength tests made from test cylinders taken at the same time. These cylinders are usually tested at 7 days and 28 days. The 7-day test gives a good early indication of the final strength since concrete reaches about 70% of its ultimate strength at that point (Figure 2-3). Concrete made with high-early strength cement attains its design strength within a very short period and is used when full strength is required as quickly as possible, primarily in items with structural properties or functions.

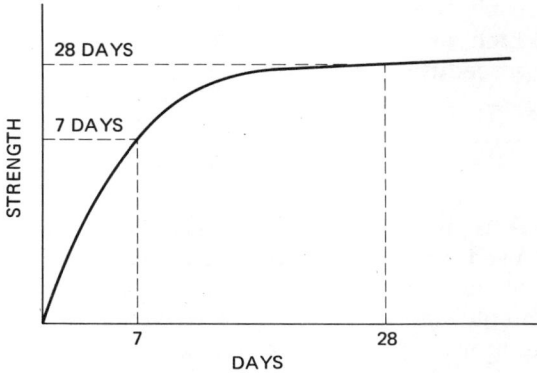

FIGURE 2-3. Concrete strength.

The quality and behavior of concrete depend on the ambient temperature when it is being placed and cured afterward. The optimum temperature range for placing concrete extends roughly between 50°F and 75°F (10°C and 24°C). If the temperature is close to or below freezing, the mixing water, and possibly also the aggregate, must be heated and the concrete must be protected and heated. Frozen concrete will not set and will tend to crumble and have no structural strength at all.

If the temperature is too high, it may be necessary to use chilled water for mixing the concrete. In setting, concrete generates heat, the heat of hydration or cementation, which can create certain problems, especially in mass concrete. This does not affect most buildings except perhaps the ones having massive raft foundations. In this case, it is important to ensure a uniform pouring schedule and to avoid the formation of *cold joints* which can result in serious structural impediments.

Concrete tends to shrink and may develop hairline cracks which do not necessarily indicate structural failure. Concrete may also expand, and in order to control this expansion and contraction joints must be provided, especially in long structures or slabs. The joints may incorporate fully separate members capable of moving relative to each other, or saw cuts filled with a resilient material, or metal strips with the cracks hopefully developing alongside them.

Since concrete is rated as a fireproof material, a concrete structure is not in need of the protection required for a steel structure. Its main disadvantage in comparison to structural steel, however, is its weight and bulk.

Concrete strength is a function of the mix which must be suited to the ingredients and is based on laboratory tests and experimentation. Although the rough proportions of concrete of different strengths are similar, the actual balancing of the ingredients depends greatly on their properties and will tend to be based on local standards.

After the concrete is placed it must be cured. The cementation process requires that a certain amount of water be supplied and the concrete has to be kept moist until it has set. Cement finishing must be done with the concrete at an intermediate stage of setting. If a separate topping is applied, the base slab must be screeded rough in order to provide a bond with the new concrete.

Concrete which is subject to exposure to the elements may receive air entraining agents to make it more durable. These substances create many tiny air bubbles in the concrete and increase its resistance to the summer and winter cycle. Air entraining agents are especially effective in sidewalks or concrete paving which are exposed to freezing and thawing. The situation is made even worse by the use of salt or calcium to melt ice and snow.

There are other additives which will either accelerate or retard the setting of concrete. The latter may be important for exposed aggregate ornamental concrete.

Other integral ingredients tend to make concrete more impervious to water seepage or capillary action.

By using certain lightweight materials as aggregates, lightweight concrete can be produced for special purposes such as roof fill, insulating concrete, and so on. Materials used for lightweight concrete are described under Aggregates. Although these concretes have low strength and are not suitable for structural purposes, they do have good thermal insulation value.

Concrete is excellent in compression but very poor in tension. Therefore, steel reinforcement is required to handle the tension in reinforced concrete.

In reinforced concrete the steel and concrete bond together and since they have very similar coefficients of expansion, they can act as a unit through changes of temperature without losing their bond.

Reinforcing for most structures is in the form of smooth or deformed straight or bent bars which come in several grades of steel, or, as used for slabs on grade, as welded wire fabric which comes in a number of standardized weights and sizes.

Special reinforcing is required for prestressed concrete. The kind used depends on whether it is pretensioned or posttensioned. Reinforcing consists of strands of high-tensile steel wires arranged in tendons which come in protective sheaths. In pretensioned concrete the tendons are stressed first and then the concrete is poured and allowed to set. The tendons, without protective sheaths, bond with the concrete. In posttensioned concrete the concrete is poured first and the tendons, in protective sheaths, are stressed after the concrete has attained design strength. In this case, there is no bond between the tendons and the concrete.

Reinforced concrete is subject to rigid codes and specifications. Most projects are carefully monitored and inspected and tests are usually required to ensure proper concrete strength and quality.

Cement and Plaster

Cement is the most important ingredient of concrete. Historically, cement is among the oldest known man-made building materials. It is known that the Assyrians used a type of cement which may have had a bituminous base. This was used to bond their bricks together which were probably similar to adobe-type brick since they were made from mud which was dried in the sun. The Egyptians used a type of burned gypsum which was also used as mortar for the pyramids. About 2000 years ago the Romans made a natural cement from volcanic materials such as pumice or puzzolan, and structures made with these materials survive today.

There are many different types of cement in use, most of which are combinations of basic materials ground to a fine powder and including limestone, clay, sand, gypsum, lime, and many other ingredients. They all have one property in common: after being mixed with water to a plastic consistency, they set into a hard substance after drying.

The most common cement is Portland cement, which derived its name when a British mason about 150 years ago experimentally discovered its similarity, when hardened, to a limestone found in the Portland region in Great Britain.

Modern cement is made by either the *dry* or *wet* process. Either process uses a large rotary kiln in which the finely ground raw materials are burned at a high temperature and are converted into clinker which afterward is ground into an extremely fine powder in a ball mill.

There are many other types of cements; for example, there are hydraulic cements (they set under water), natural cements, gypsum cements (they tend to set slower due to the addition of lime which acts as a retarder), aluminous cements containing aluminum ores, masonry cements intended for mortars, and many others.

Cement mixed with sand, lime, or other ingredients produces

mortars which are used for bonding masonry together. It is also used in making grouts and different types of cement finishes that are used on floors and vertical surfaces.

Plasters are made with gypsum and are used for hardened wall and ceiling finishes which can be painted, or they can be used in drywall work where a hardened layer of gypsum between special covers forms a sandwich which is used as a finish which, once the joints are taped, is ready to paint or paper.

The fire resistance of gypsum is one of its most important qualities. For this reason plaster and drywall are sometimes used to cover structural steel in columns or beams. Plaster, being one of the "wet" trades, is much messier than drywall and takes longer to install because each of the different coats must get enough time to dry before starting on the next one. The work may be either three- or two-coat work. In the first case, there is a scratch coat, then a brown coat, and finally a white coat or finish coat. In two-coat work the last two are combined. Other types of plaster are various cement plasters and hard facings like Keene's cement. Drywall also produces a bit of a mess when the joints are sanded and result in a fair amount of dust. Since the only waiting time is the time required for letting the joints dry out after taping, the total finishing time is considerably less than for plaster.

Aggregate

As one of the basic components of concrete, aggregates often tend to be locally available materials for primarily economic reasons. Often aggregates are granular materials such as sand or gravel, but for the coarse aggregate limestone or other suitable rocks may be used. In areas where no natural suitable materials are found the aggregates must be imported and, of course, the cost of the concrete is higher.

Since the aggregates must be graded in different sizes, natural deposits of uniform size, such as sand from a desert region, would not be very useful in setting up a concrete plant. It is also worth noting that certain minerals could have an adverse reaction with reinforcing steel if used in concrete.

For lightweight concretes, different types of aggregates are used including materials like slag, cinders, fly ash, perlite, haydite, vermiculite, zonolite, volcanic stones, shale, slate, and many others. Some of these have corrosive components and must not be used in contact with metals.

Water

Water used in concrete must have a certain quality to be suitable for mixing water. Hardness is not necessarily a problem, but sea water would not be suitable. Perhaps the simplest rule to remember is that if the water is potable then it is generally suitable for mixing concrete.

As mentioned previously, concrete placed under winter conditions requires heated mixing water (perhaps the aggregate should also be heated). Conversely, in extreme heat mixing water for concrete may have to be chilled. The amount of water to be mixed must be carefully controlled because too much water weakens the concrete. It is advisable to have good site inspectors whenever concrete is being placed because there is a tendency for the men to add more water to make the placing easier by having a soupier mix.

MASONRY

Masonry units are generally of a modular nature. The materials include loadbearing and nonloadbearing units, mortar, masonry hardware, and accessories.

The historically oldest and still most common building material is brick, and there are many types, colors, patterns, and modules. In the beginning brick, which originated in countries having warm climates, was made from mud, at times reinforced with straw or reeds, and left to dry in the sun and bake to a hard consistency. Mortar also consisted of mud mixed with sand and was dried under the sun. This method in the form of adobe brick has survived in Africa and South America until the present and is the ancestor of modern masonry.

Brick comes essentially in two main categories; burnt and unburnt. Within each category there are many varieties.

The majority of brick used is burnt clay brick. For economic reasons, the materials are usually native to the place of manufacture. And depending on their composition and the manufacturing process, the materials affect the strength, color, texture, and so on, of the brick produced from them. Brick may be molded or extruded, solid or cored, smooth or textured.

Brick sizes are standardized across the continent.

Most burnt clay brick is loadbearing, but brick walls can also be reinforced by special joint reinforcing inserted in the mortar at specified course intervals or by interlocking headers in the brick pattern. Other types of brick include cement brick, silica lime brick, and firebrick for lining flues and combustion chambers. Some brick is made for special purposes in chemical plants and requires special compatible mortar.

There are different patterns in which exposed brick is laid. Some of these *bonds* are shown in Figures 2-4 through 2-8. Finishing of the joints also has a bearing on their durability as well as on the visual aspects of the wall, since it can give it a special textured appearance depending on the joint pattern used. (See Figure 2-9.)

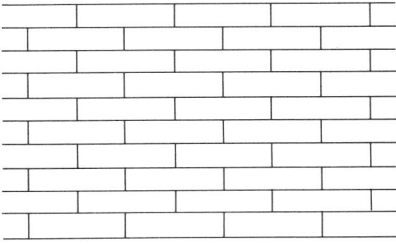

FIGURE 2-4. Running bond.

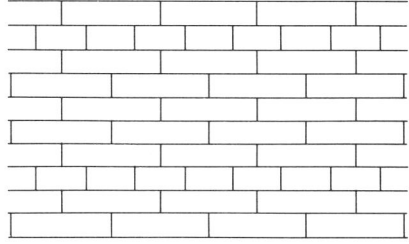

FIGURE 2-5. Common bond.

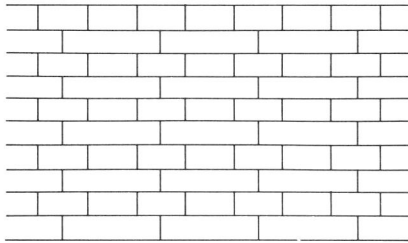

FIGURE 2-6. English bond.

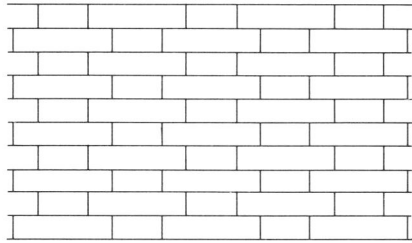

FIGURE 2-7. Flemish bond.

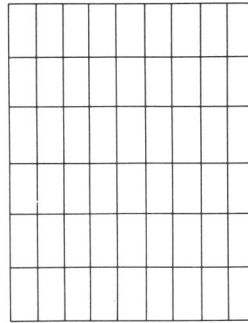

STRETCHERS   HEADERS

FIGURE 2-8. Stack bond.

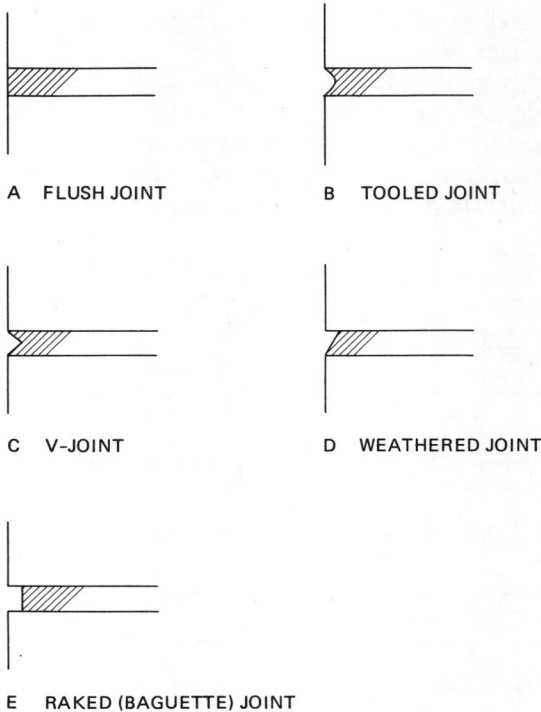

A   FLUSH JOINT   B   TOOLED JOINT

C   V-JOINT   D   WEATHERED JOINT

E   RAKED (BAGUETTE) JOINT

FIGURE 2-9. Brick Joints.

The thickness of brick walls is usually tied into the height of the walls and this is governed in most cases by local codes. When brick is used as facing only, such as in brick-veneer walls, the code may require pressure-relief angles at specified vertical distances to prevent possible failure and collapse of the facing.

After the brick is laid, the brick is washed down and cleaned off with muriatic acid. A silicone coat is sometimes applied to dampproof the exterior masonry, but this has its limitations.

Brick consists mainly of different types of clay, with other ingredients added which affect color, texture, water absorption, strength, and so on. After being mixed with water in special pug mills, the mixture is either extruded through special molds and cut into individual units or it is molded into special forms. After being dried and fired in kilns under strictly controlled conditions, the brick is allowed to cool, again under controlled conditions.

Glazed brick, in which a specially applied glaze is fired to a vitrified consistency, comes in a number of types and finishes. It is extremely expensive and is usually custom-made for specific projects. It is very important to order a sufficient quantity of stretchers, headers, quoins, and special shapes from the same dye lot in order to ensure that the color is uniform throughout and because of the long lead time required for additional orders.

Roman and Norman brick are usually longer or thinner and vary in size only.

Split-face brick, including Roman and Norman brick sizes, is a double brick having a thin neck between the units. When the neck is broken, two bricks are produced, each having a rough-textured longitudinal band which creates a special visual effect in the wall when used as face brick.

Another specialty brick is radial brick which is made with a curvature and is used for stacks and chimneys. Radial brick is not used very often today since many chimneys, especially high ones, are made with reinforced concrete and are lined with special refractory materials.

Concrete block which is generally loadbearing, although made of concrete, is generally classified under masonry. The blocks come in a number of standard modules, primarily 8 in. × 16 in. (20 cm × 40 cm) and usually in thicknesses of 4 in., 6 in., 8 in., 10 in., 12 in., (10 cm, 15 cm, 20 cm, 25 cm, 30 cm). The blocks, depending on the manufacturer, have two or three vertical hollow cores. There are also solid block, suitable for piers and pilasters, but columns are sometimes also made by inserting reinforcing bars through the cores and filling the spaces with concrete. Block walls can be reinforced by means of joint reinforcing, usually at every third course, and placing the reinforcing in the mortar before laying the next course.

Various other concrete blocks are made to the same dimensions but with materials such as slag, cinders, haydite, and so on. They are generally lightweight blocks, are nonloadbearing, and have certain thermal and sound insulation properties.

In addition to the regular block used in walls and partitions, there are varieties of ornamental concrete block for use in decorative open-work patterns. These blocks are usually square in shape and the patterns vary with the manufacturers. Other specialty block include chimney flue bases with provision for a clean-out and used to support prefabricated chimneys. Other specialty block are made for lintels over openings and for jambs for windows and doors, and there are half-block to close off gaps in the serrated pattern of staggered blockwork. Two-inch (5-cm) soaps are made as liners but they are not commonly used.

Concrete block are usually made of a higher strength concrete and sometimes high-early strength cement is used in them. Some of the better block are steam-cured and a finer aggregate is used in them to give a smoother texture.

Various masonry mortars (especially cement mortars) are used with concrete block.

Terra cotta, which is also burnt clay material, comes either as load-bearing or nonloadbearing. There are many different shapes and modules depending on whether the terra cotta is to be used in partitions, as wall lining, bonded with face brick, or for fireproofing steel beams or columns. The cellular structure of terra cotta has good thermal insulating properties and its texture is suitable for direct plaster application, but it is fragile and more subject to breakage than are concrete block.

Gypsum block, which comes in a number of thicknesses, are not too widely used because of their high proportion of breakage. They must be used as a plaster base and because of their dusting cannot be left exposed.

Siporex blocks, which can be sawed and nailed, were more widely used a number of years ago. Extremely popular for years after being introduced here on this continent from Scandinavia, they have lost some of their appeal possibly because they are expensive to manufacture.

Glazed and unglazed tile, which come in modules differing in size from terra cotta, are often used as linings and dadoes in corridors, schools, and hospitals.

Glazed tile must be ordered with all the special shapes which allow for corners, bases, jambs, lintels, sills, bullnose transition pieces, and so on. The composition of the raw material and the firing of the glaze influence their quality. They can also cause crazing, which is a distinct pattern of irregular lines that can appear on their surface after some time. (The problem is similar to the one in ceramic tile.) These block must be handled very carefully during laying to avoid damaging the glazed surfaces.

Special materials include flue liners and refractory materials such as firebrick.

Glass block, although in theory not a masonry material, are usually included because they are often installed by the masonry contractor. Because of their high expansion it is important to set them in a manner which permits lateral and vertical movement to take care of this expansion. There is also the danger of the joints opening up if the glass block panels are too large or do not have proper expansion joints.

Stonework also belongs under masonry. This includes natural materials such as limestone, sandstone, granite, marble, travertine, slate, fieldstone, artificial or precast stone, and other manufactured stone.

There may be an overlap with tilework. The distinction is usually the criterion of whether the units are bonded to each other or surface-applied, which is a function of their thickness.

Stonework can be set in different patterns. Fieldstone or rubble stone is set in a completely irregular pattern. Cut or hewn stone is set in random or coursed ashlar. Stonework can have a rough-textured finish or a smooth polished finish.

Mortar for stonework must be carefully chosen for each type of stone, because of the difference of porosity and moisture absorption of different stones, and the possibility of adverse chemical interaction between the mortar and certain minerals in the stones.

Openings in masonry work often require special provisions for lintels, sills and jambs. Lintels are required to support the overlying masonry and consist either of steel angles, precast concrete, or cast-in-place concrete using forms or special lintel blocks. In brickwork, curved arches are sometimes formed over temporary curved forms. Sills may be formed of brick, but they are usually made of precast concrete or cut from stone. They must be shaped to suit the window or door. Jambs, if in a wall other than brick which can be cut easily with a trowel, either have special block or are cut from regular block with a masonry saw to provide neatly finished units.

Stone paving may come under masonry, but it is not always set in mortar. It may be placed on a sand bed to permit it to move freely against frost action.

Stucco, although often applied by plastering contractors, is often included with masonry because of its close relation to it. Stucco may be applied directly on masonry units or over metal lath which is more durable. Although it can be two-coat work, a good stucco job requires three coats with the final coat determining both color and texture. It can be finished smooth or with a rough and/or patterned texture. Sometimes pieces of brick or stone are intentionally left sticking out to provide a special visual effect.

METALS

A number of metals play major roles in the building industry. Foremost in line are various grades of steel which are primarily used for structural purposes and can be subdivided into different categories such as those indicated in the following listing:

   A. Structural Steel:
      (1) Steel framing
      (2) Steel decks and pans
   B. Concrete Reinforcement:
      (1) Reinforcing bars
      (2) Welded wire fabric
      (3) Prestressing tendons

Steel consists basically of iron with a fraction of a percent of carbon added. Varying the carbon content results in other alloy composition, for example, wrought iron and cast iron. Different chemicals added to the steel to change the alloy composition affect some of its properties such as strength, ductility, resistance to fatigue, and so on. Adding nickel and chromium results in stainless steels. Since steel exposed to moisture will rust, the steel must be protected. There are, however, steels that are designed for permanent exposure and that acquire a coat of protective rust which actually becomes a visual feature.

Structural steel is used by far the most in buildings. There are different grades of steel not only for structural members but also for ancillary items such as rivets, bolts, steel ropes, and so forth.

Steel decks come in a number of shapes and gauges, depending on whether they have to support floor or roof loads. They can be fully cellular decks, which are the strongest, or simple decks having raised ribs. For reinforced slabs over steel there are also flat sheets having raised V-joint stiffeners.

Reinforcing steel comes in a number of grades and as regular or deformed bars. Deformed bars have a serrated surface to provide more bond with the concrete. There is a series of standard sizes ranging from $\frac{1}{4}$ in. to $2\frac{1}{4}$ in. (6 mm to 57 mm) in diameter.

Welded wire fabric comes in different grid modules and variable wire sizes forming the fabric. One of the standard sizes often used is 6×6 ($\frac{6}{6}$) which indicates No. 6 wires in both directions spaced to form 6-in. squares.

Prestressing steel tendons are made from special high-tensile steels. When used for posttensioning, the tendons must be made to measure with the anchorages fastened to the tendons. When used for pretensioning, the excess length of the tendons is cut off.

Other structural uses are for pans for ribbed and waffle floors. The trend is to make them out of reusable plastic or single-purpose and single-use cardboard.

Steel has extensive use in other building components and elements such as steel sash, miscellaneous ironwork, piping, conduits, and mechanical and electrical equipment. In addition, sheet metal and sheet steel are used for sidings which may be prefinished in various colors or which may have porcelain-enamel or baked-on paint finishes, equipment housings, cabinets, lockers, toilet partitions, doors and frames, roofing, and roof flashings and copings.

Stainless steel finds applications in laboratory, kitchen, and cafeteria equipment, curtain walls, storefronts, doors and frames, hardware, and other specialty uses.

One of the more exotic versions of steel destined for widespread use in the future, is dual-phase steel. It is steel that will be capable of being deformed by tension or torsion, and gain strength as it is worked on. Before it finds use to any extent on construction sites, it is more likely to be used extensively in manufacturing of machines and automobiles. Because of its lighter weight and energy-saving properties it will become extremely popular as soon as it reaches commercial viability.

The second most used metal is aluminum. It comes in various alloy compositions since pure aluminum is too soft and weak. It is used for storefronts, doors and frames, windows, sidings, roofing, copings, flashings, electrical conduits, miscellaneous equipment and ornamental work, and hardware. Because of its soft composition, it requires special finishing. Most items have anodized finishes. Aluminum comes in cast or wrought alloy composition. There are various treatments and processes to improve properties such as durability, corrosion resistance, strength, finish, and so on.

A number of other metals are used in construction but to a lesser degree. Brass, bronze, and copper are used for ornamental work and various other purposes, including piping, wiring, and roofing and flashing.

Lead is used for X-ray protection, paints, roof metals in salt-water atmospheres, drains for corrosive fluids, etc. It is not recommended for water piping for human use because of its toxicity.

Nickel and chromium are used for finishing of metals and in the composition of stainless steels.

Most steels can be welded, but it is important to use the right method for each material. There are a number of welding methods, including electric arc welding and gas welding and variations of either. Wrought iron and cast iron are not suitable for welding. Aluminum alloys are difficult to weld and not all of them are weldable.

Some other metals such as copper, tin, etc. cannot be welded but they can be joined by soldering which does not develop the same strength.

Since welding and soldering involve the application of heat, it must be kept in mind that they may destroy the effect of heat treatment which the metals to be joined may have undergone, and can, therefore, have adverse effects. It may also be necessary to anneal the metals again if this is important for the ultimate use. In some cases, preheating is done as a precautionary measure to avoid premature failure.

Since most metals are subject to corrosion, in most cases they require protection if they are to be exposed. This protection can be in the form of either special finishes or anti-corrosive paints which either form an airtight protective covering or react with the surface to form a protective film. An alternative, used primarily in mechanical and piping installations, is cathodic protection in which a countercurrent is set up to resist and neutralize the currents set up by galvanic action between uneven metals in contact exposed to moisture.

ASBESTOS

Asbestos is a mineral that has a fibrous structure. Its main distinction is its fireproof nature. The main deposits occur in Canada, South Africa, and Russia's Ural Mountains, but there are minor deposits in the United States. The ancient Romans used asbestos for burial cloths to preserve the ashes after cremation.

In addition to its fireproof qualities, asbestos is also an excellent electric insulator and is used as such.

In building construction asbestos is utilized extensively in combination with cement. Asbestos cement in rigid sheet form is used for firebreaks on partitions and in ceiling spaces. Corrugated sheets are used as sidings and on sloping roofs. For the latter there are also asbestos shingles. These must be predrilled to avoid splitting them, for they are very brittle and do not have much elasticity. Special sections are used for rigid hollow roof decks.

As a plastic slurry, asbestos is used extensively for insulation and fireproofing structural steel and decks. Being blown under pressure, it can be applied in hard-to-reach locations and follows irregular contours. Because of its dusting it must never be used in an exposed location or in ceiling spaces which serve as plenums for air conditioning or ventilation. Asbestos dust is a serious health hazard. For this reason protective masks must be worn when working with asbestos materials. When asbestos board is used in an exposed location it should be painted to eliminate or reduce the dust problem.

PLASTICS

Plastics are artificial products of modern chemistry. Many of them are derived from petrochemicals. Plastics can be rigid or flexible, and they come in bulk or in sheet form. They usually combine high strength with low weight. Some may have a tendency to deteriorate with time; others are attacked by various chemicals or pollutants in the atmosphere or other extraneous agents. A number of plastics support combustion and give off highly toxic gases.

In construction they are used primarily in plastic laminates used for built-in items, furniture, paneling, counter tops, vanities, dadoes, protective coverings, and wall finishes. Other plastics include epoxies that are used for flooring, roofing, and exposed finishes. Still other plastics are used for insulation (styrenes and foamed plastics). Clear transparent acrylics are used in skylights and vision panels. For a number of years plastics have been used in making complete modular units such as bathrooms for hotels or motels, modular units for kitchens, and other prefabricated modular units.

A more exotic use of plastic which is still in the experimental stage is in the production of dome-shaped prefabricated or cast-in-place mushroom-shaped housing modules. Some scientists predict, with tongue-in-cheek, that with the trend of using plastics in housing, the house of the future will be washed down with a hose instead of being cleaned with a vacuum cleaner.

There is a definite and noticeable tendency of plastics to displace and replace more and more items usually made of metals. This is partly due to the intention of saving weight and partly to saving costs. In addition, plastics do not deteriorate the way metals may corrode or rust, although some plastics are subject to other forms of deterioration.

There are many types of plastics, but they are grouped under two main categories: thermosetting plastics and thermoplastics. Thermosetting plastics may start soft but they harden after continued heating. Thermoplastics can be softened by heating and rehardened by cooling.

The first true plastic was Celluloid, which was invented about 1870. This was followed about 1910 by Bakelite. From then on development of plastics followed almost on an exponential scale.

Plastics are made by a number of methods. Most plastics are

molded, mainly by injection or extrusion molding, compression or blow molding, or several other molding methods. They can also be made by casting, by laminating or coating, or by a number of other manufacturing methods.

Some plastics display a built-in memory. If deformed, they will return to their original shapes.

Plastics may be reinforced for additional strength. Among the materials used for reinforcing, glass fibers are high on the list because of their ability to withstand heat and because of their relative chemical inertness.

Some plastics such as foamed-in-place types used for insulation have single-purpose applications.

Laminated products are often made under a combination of high temperature and pressure. When they are applied to other surfaces, it is imperative that the right type of adhesive under the proper temperature and humidity conditions is used.

RUBBER

Rubber is produced from an elastic substance which is extracted from various tropical plants and trees. In its natural form, called *latex*, its use goes back to ancient times in countries in Central and South America from where early explorers and conquerors brought it back to Europe.

Once rubber is vulcanized (vulcanizing consists of various treatments combining different chemicals with heat and pressure), it will not become soft or sticky under heat or become rigid under cold.

The construction industry mostly uses synthetic rubbers such as neoprene or butyl but there are many natural or artificial rubbers. The properties of these which are of commercial importance include, among others:

- Elasticity/energy absorption
- Compressibility/elongation/strength
- Hardness/softness
- Expansion/contraction
- Resistance to:
  aging
  friction/wear
  heat/cold
  sunlight
  passage of gases
  electricity

Rubber can be laminated and it is sometimes sandwiched with metal for special uses. It is used to cover wiring and in vibration pads and shock absorbers, washers and gaskets, drive belts, hoses, special roofing, industrial doors, waterproofing, and floor coverings, but its main use is in tires for mobile construction equipment.

BITUMINOUS PRODUCTS

Bituminous materials are mineral substances of organic origin consisting primarily of hydrocarbons. Some of these, like asphalt and tar, occur in natural lakes which often contain the preserved remains of prehistoric fauna which happened to fall into them.

The bituminous products used in construction are mostly derivatives from the petrochemical industry, such as asphalt which results from distillation of petroleum. Asphalt, pitch, tar, and other related products are used in paving, roofing, asphalt-impregnated building boards, papers, felts, dampproofing and waterproofing compounds, foundation coatings, sealers, and flooring materials.

In their main applications of paving and roofing the economy and efficiency of bituminous products have been proven by the fact that they have endured unchanged and unchallenged by substitutes.

PAINTS

Paints are mixtures of basic white with or without colored pigments, binders, solvents, fillers, and driers. There are different types of paints which may be made with a water, oil, latex, alkyd, or other base or composition. House paints are generally coarser than artists' paints.

The pigments in paint provide the protection of the painted surface as well as the color, and they are held together by binders. The solvents penetrate porous surfaces and act as lubricants while the driers speed up drying and harden the finished surface. The fillers help to make pigments lighter or heavier. Because they are less expensive than pigments they help to reduce the cost of paints.

In building construction, paints are used for varied applications. Surfaces and objects finished on site may be painted with either flat or enamel paint. Enamel provides a harder and more durable surface than flat paint does. Wood may be stained or varnished and flooring may have special sealers applied. Manufactured items may have baked-on enamel finishes. Special paints are made for masonry and cement. Other paints are anti-corrosive, anti-static, and heat-reflective. Special plastic paints result in finishes which have a rough surface and may be multi-tone in appearance. They are useful in areas of high wear such as passages, staircases, storage areas, and workshops.

Because of the trend toward more prefabrication and prefinishing, many materials come prepainted or with integral finishes.

In field painting (on the job painting as opposed to prepainting in the factory) one of the most important points to keep in mind is to use the right primer and not to skimp on quality because the primer governs the adhesion and lasting quality of the complete paint job.

Painting is usually two- or three-coat work. In the former case, the base coat may be tinted to give better coverage for the finish coat. For outdoor work, three-coat work is recommended and the right type of paint must be used. Since gypsum board generally uses more paint than do hard plaster surfaces, allowance must be made for this. Sufficient time must be given for each coat of paint to dry and to set. Surfaces to be painted must be clean, dry, and free of oil or grease. When old surfaces are being painted over, they should be thoroughly cleaned, old flaky paint should be burned or scraped off, rust spots should be removed, and holes and cracks should be repaired before any new paint is applied.

FUELS

Fuels are substances, mostly of organic origin, which can be burned in air at a fairly rapid rate and generate heat or power that can be utilized to drive equipment or machinery or to heat buildings.

Fuels are classified as solid, liquid, or gaseous.

Solid fuels include coal, coke, charcoal, peat, and wood.

Liquid fuels include petroleum products, alcohol, and colloidal substances.

Gaseous fuels include manufactured gases, natural gas, and pure molecular combustible gases.

Most fuels, being of an organic nature, consist primarily of carbon and hydrogen. They also contain oxygen, nitrogen, and other elements such as phosphorus and sulfur. In addition, they may contain nonvolatile components which result in ash, the residue after combustion.

The calorific value of a fuel is the measure of the heating capacity per pound of fuel, or calories per kilogram. This measurement is expressed in British thermal units (btu's) and is the amount of heat required to raise the temperature of 1 lb of water 1°F. An alternative definition is to express the ratio of the weight of water evaporated per unit weight of fuel burned, assuming no heat loss in the process of evaporation.

Coal is one of the most important fuels, especially in view of the shrinking petroleum reserves and foreign dependency on petroleum. By contrast, coal reserves on the North American continent are considered sufficient for several hundred years.

Coal, which is a mineral carbonization product of vegetable matter over a long period of time, is classified by ranks in terms of its free carbon content and its geologic age.

The best grade of coal is anthracite, which consists primarily of free carbon. Anthracite is subclassified into several types which vary primarily in hardness.

The next grade of coal is bituminous, which ranges from low to high volatile grades.

The next grade is subbituminous coal, which is also divided into several subgroups.

The lowest rank of coal consists of the lignites, which are either black or brown. They represent an early stage in the development of coal from wood.

The quality of coal often depends on the type and degree of impurities found in it. One of the major problems is caused by the occurrence of sulfur, either as iron pyrite or some other sulfate combined with carbon. The main objection to sulfur is that on combustion it forms highly corrosive acids such as sulfuric acid. This means that coal that has a high sulfur content requires special treatment, which, of course, raises the cost. Certain other impurities result in a high ash content which can be objectionable for certain commercial or industrial uses. Some of these impurities can be eliminated, but any extra treatment may be costly enough to make use of this coal uneconomical for some purposes.

The increase in the cost of petroleum during the past few years, however, has turned the focus back on coal, which even with its impurities has turned from an ugly duckling into a swan.

Coke and charcoal are the result of heating coal and wood in enclosed containers (coke ovens) and distilling the volatile components. Coke is also obtained as a byproduct in the production of manufactured gas. It is used extensively in the steel industry in blast furnaces.

Charcoal is produced by partially burning or oxidizing wood. Charcoal has a very low density and high porosity and because of its adsorptive properties it has extensive uses other than as a fuel, for example, in various types of industrial filters.

Peat consists mainly of decomposed vegetable matter and is generally found in marshland. It is comparable to coal at an early stage of carbonization. Its high content of moisture renders it impractical for industrial purposes. Peat is used extensively in Northern European countries, but its use on the North American continent has been limited.

Wood is not of great significance as a fuel. Its use is generally restricted to local conditions and often it is only the result of waste products from other industries such as lumber manufacture. Its thermal value often is a function of its resin contents, which tend to be higher in softwoods; it also is inversely proportional to its moisture content.

Petroleum products range from fuel oils, which are on the lower scale in the hierarchy of the processing of crude oil, to gasoline which requires a great deal more refining. They vary greatly in calorific value as well as in flash point, which is the temperature at which they will ignite. In fact, certain heavy oils require preheating before combustion, but once they burn, they are also more difficult to extinguish. Oil fires are among the most difficult fires to combat.

Fuel oils generally do not leave any residue and if they are mixed with air in the proper proportion, they also result in comparatively clean flue or exhaust gases, which can be cleaned even more effectively by using special scrubbers or electronic precipitators.

Gasoline, used primarily in passenger automobiles, is one of the most processed items in this order, and it is often modified by admixtures of various substances for specific purposes. It is interesting to note one development that has come full circle. In the 1930s lead was added to gasoline in order to eliminate knocking in automobile engines. The knocking was caused by premature ignition of the fuel mixture in the cylinders. In the late 1970s the trend is toward unleaded fuels because of the damage the lead does to other components of the automotive engine. Incidentally, the catalytic converter used by some car manufacturers cannot use leaded fuels. It is essentially an after-burner to dispose of certain unburned components in the exhaust gases and since, in this capacity, it generates extremely high temperatures, it has been attacked by its critics as unsafe and a fire hazard.

For commercial and industrial vehicles, such as buses, trucks, and construction equipment, gasoline is too expensive. Therefore, diesel fuels, kerosene, and other less refined fuel oils are used.

Alcohol, which can be produced from many materials (often the waste products of industrial processes), is not used extensively on the North American continent as a fuel, but, as a result of the fuel crisis, there may be an increase in its importance, specifically in the form of methanol or wood alcohol, which can be produced by distilling wood. In tropical countries bagasse, or empty sugar cane stalks, are used extensively, and in more temperate countries sugar beet is finding wider use.

Another less conventional fuel is gasohol which is a hybrid of gasoline and alcohol. It was developed to provide an alternative fuel in case of serious gas shortages or depletion. It has found limited use primarily in agricultural regions where the necessary products can be found to produce the alcoholic components of gasohol.

A number of colloidal fuels have specialized use. They consist usually of a combination of a pulverized solid fuel mixed with a fuel oil to form a semi-liquid or gelatinous substance.

Manufactured gas is produced primarily from distilling coal products, but it is also produced as a byproduct from the cracking of petroleum products or in steel manufacture, such as blast furnace gases. There are different grades, which are classified according to their thermal values.

On the North American continent a lot of natural gas is found, but there is not enough to meet the ever-increasing demand. Liquified gas can be transported in compressed form and then deliquified. In the near future natural gas may be imported from distant places such as Northern Africa or Russia where it occurs in abundance but is mostly wasted.

Pure molecular gases such as hydrogen, oxygen, and so forth, have specialized uses and are of no particular significance in construction except in welding and cutting.

Both liquid and gaseous fuels are easily transported by pipeline. The oil crisis of the early 1970s has emphasized the importance of northern oil and gas reserves, and the various schemes for linking Alaskan deposits with their southern destinations have proven to be lengthy and contentious issues.

3

Building Components and Construction Methods

FOUNDATIONS

Every building or structure rests on a foundation. The purpose of the foundation is to support the weight of the building and its contents and to transmit these loads to the underlying geological strata in a manner to ensure that there is either no or only nominal subsidence or settlement of the building.

Foundation design is based on a number of different considerations. Some of these limit the choice to specific methods; others permit alternatives which can be evaluated on the basis of optimum economic impact.

Among aspects to be considered are the following:

A. Size and weight of the building
B. Static and dynamic loads of the contents
C. Nature and consistency of the underlying geological strata
D. Soil bearing pressure applicable
E. Ground water table and characteristics
F. Special design conditions such as wind and earthquake loads, etc.

Generally, the first consideration for the average building is whether it is to sit on rock or on soil. Most types of rock have such high load capacity that they can be considered incompressible for all practical purposes. This means that a building erected on this rock will not settle. There are, however, exceptions to these assumptions and very large high-rise buildings in particular require careful analysis in this respect.

Different soils have different loadbearing capacities which depend on the composition, compaction, cohesion, porosity, and moisture content and retention of the soil. A building resting on these soils will settle and it is important not only to try to minimize this settlement but also to ensure that it is uniform across the whole building. Otherwise, one portion will settle more than another one, and the resulting differential stresses could cause serious structural damage, and in extreme cases, failure.

Foundations can be divided along broad lines into the following groups:

A. Spread footings
B. Floating foundations
C. Piles and caissons
D. Anchored foundations

Spread Footings

Spread footings are the most common type of foundation. Spread footings transmit and distribute concentrated loads from columns or walls over a larger area of the underlying soil. It is obvious that the larger the footing, the lower the bearing pressure exerted on the soil underneath. Consequently, the settlement is also lower. In order to effect a gradual transfer of loads, pedestals may be placed between the columns and the footings (Figures 3-1 and 3-2).

If the building rests on rock, only nominal footings may be required. Broken or weathered rock may be treated as if it were a soil having a higher bearing capacity.

Floating Foundations

Under certain conditions, spread footings are not adequate or would, in theory, cover the complete building area. In such cases, the building may be designed to rest on what is generally known as a *raft foundation.* This is a very heavy slab which is often strengthened by inverted beams along grid lines, which in turn support the walls and columns. A building on a raft foundation will settle uniformly, but special provisions will have to be made to allow for movement in the utility services to the building. Sewer, water, gas lines, and electrical conduits must be protected by means of adequate safeguards

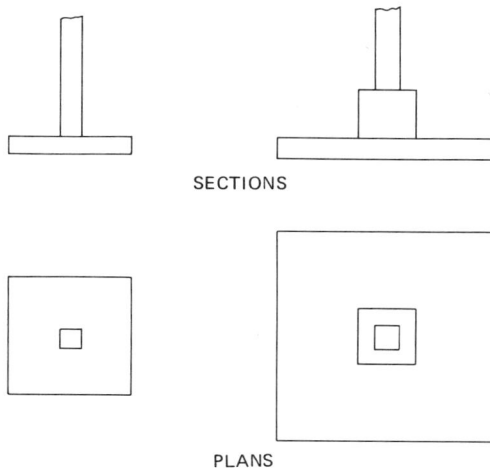

SECTIONS

PLANS

FIGURE 3-1. Column spread footings.

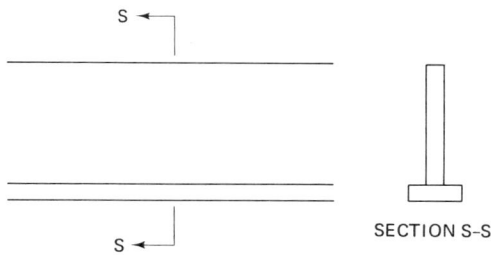

SECTION S-S

FIGURE 3-2. Wall footings.

against damage due to differential settlement of the raft (Figures 3-3 and 3-4).

Raft foundations are relatively rare.

Piles and Caissons

A building resting on unstable soil, or at a vertical distance from solid rock which would tend to make the use of piers to rock uneconomical, or a building which exerts loads which would require excessively large spread footings, may be supported on piles or caissons.

Piles are subdivided into two main categories: Figures 3-5 and 3-6 (A. Bearing Piles. B. Friction Piles).

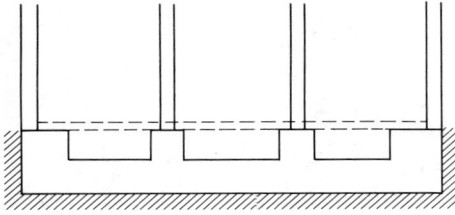

FIGURE 3-3. Raft foundation.

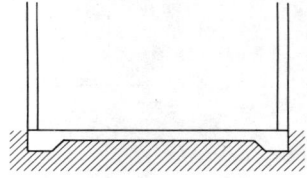

FIGURE 3-4. Mat foundation or floating foundation.

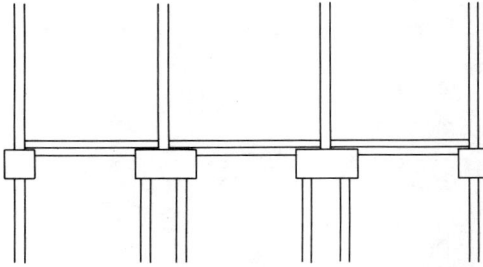

FIGURE 3-5. Pile foundation.

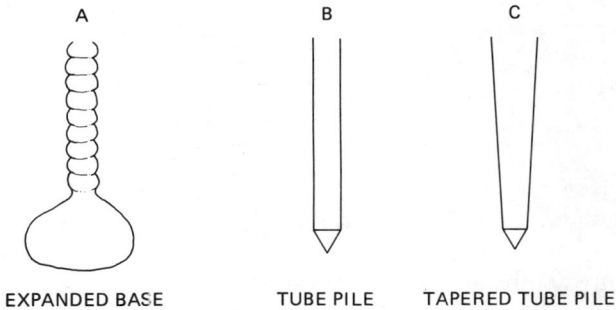

EXPANDED BASE   TUBE PILE   TAPERED TUBE PILE

FIGURE 3-6. Types of Piles.

Bearing Piles

Bearing piles are used when there is a definite and limited distance to rock or other surface having a high enough loadbearing capacity. A number of different types of piles are used as bearing piles and there are several methods of implanting them. Some of these piles are described in the following paragraphs.

H-Beam Piles. These consist of structural sections which are driven into the ground until they reach solid rock or refusal. With these piles, as with other types, it is possible to hit a boulder and think that solid bottom has been reached. When it has been determined that it is a boulder (the experienced contractor can often tell from the reaction of the machine and/or from the sound), driving must be resumed or additional piles must be driven to achieve the total design load bearing capacity.

Tube Piles. These consist of a steel shell which is driven down to refusal and then filled with concrete. Tube piles usually are reinforced by means of a prefabricated cage of reinforcing steel which is lowered into the casing before concreting. The shells may be straight or tapered and they are often driven with a special mandrel inserted into the shell during driving and then they are withdrawn before concreting.

Expanded Base Piles. A heavy tube is first driven down to refusal and then withdrawn slightly. Concrete is forced down the tube and expands at the bottom to form a bulb-shaped base of dense concrete. A mandrel is used to compact the concrete in the tube. The tube is now gradually withdrawn and concrete is poured into the tube and compacted inside the tube to form a shaft until cut-off elevation is reached. Expanded base piles cannot be reinforced in the conventional manner. Although they save the cost of permanent casings, they have some disadvantages which may offset the savings. It is not possible to tell whether the pile will turn out to be uniform and will not "neck," i.e., have a reduced cross section somewhere along the shaft that would be beyond permissible limits.

Precast Concrete Piles. These consist of reinforced concrete and are driven to refusal in the conventional manner. When the length of the piles is known, the contractor may decide to cast the piles right on the site and set up a casting yard for this purpose. A casting yard consists of a number of forms which are used over and over to produce the required piles. The following points are noteworthy. The concrete must be cured until it reaches not necessarily full design strength but until it is strong enough to permit the piles to be lifted out of the forms without cracking. Although some piles will be strong enough in compression, they will not be able to take the bending moment from their own weight if they are improperly lifted from the forms, for example, when they are lifted from one end only.

Some of the reinforced concrete piles on the market come in short lengths and are combined by means of specially patented interlocking joints. Although they are practical in some respects, they are not economically competitive.

Timber Piles. These are seldom used for buildings any more, except perhaps for temporary structures. If they are driven in soil whose moisture content varies, they are likely to decay. For this reason they must be well waterproofed, usually by the osmose process or by creosoting, but even these processes have their limitations.

Piles may be driven in clusters, either to attain a certain load capacity or to provide additional stability. Stability is sometimes attained by driving piles at an angle. Batter piles are an example. Driving piles in clusters applies primarily to H-piles which do not displace too much material. Expanded base piles have to be spaced much farther apart and are not as suitable in this respect. Since tube piles displace a fair amount of material, there is some interaction between piles in a cluster. This interaction is usually allowed for in the design.

Under certain conditions and depending on the material encountered, piles must be drilled instead of driven. This is especially true when there are numerous boulders which will stop a pile before it reaches proper bearing. Drilled piles take longer and are more expensive than driven piles.

In tight spaces and when there is not enough headroom for the pile driver, piles may have to be jacked into place, often in sections. These intrusion piles are extremely expensive and take a lot of extra time to implant.

Friction Piles

In some locations the depth of the underlying material precludes the use of bearing piles and friction piles must be used instead. In most cases, friction piles consist of a shell, usually corrugated and often tapered, which is driven until the skin friction of the shell causes a specified resistance. At this point the pile is reinforced and concreted. Friction piles are usually fairly long and require a proportionately longer time to implant.

When they are used in a cluster, friction piles must be spaced far enough apart to ensure that the skin friction of the piles that have already been driven is not reduced by the disturbance of the adjoining material.

Caissons

In the case of extremely heavy loads, caissons may be used instead of piles. Caissons consist of large built-up tubes, which may be made up of flanged sections joined together. Caissons are sometimes driven, but more often they are sunk by excavating the material from inside the caissons. This can be done by means of a clamshell or by using a rotary drill-type blade arrangement. When the caisson reaches the solid rock face, a reinforcing steel cage is inserted and the caisson is concreted. Caissons may run into special problems if they are sunk in materials such as plastic clay or quicksand. For example, they may fill up faster from the bottom than they can be excavated. Thus before they are used, accurate information on subsoil conditions should be obtained.

Anchored Foundations

In some rather rare cases a building may be erected near a body of water. The foundation may be fairly deep below a permanent head of water and subject to buoyancy. Since the basement would have to be absolutely waterproof, the buoyancy would create a tremendous uplift and there would be a tendency to lift the building out of the ground. If the foundation rests on rock, the foundation must be anchored to the rock in order to prevent the building from floating up. In such a case, the basement floor may be designed as an inverted slab. The water pressure becomes an inverted floor load. When this foundation is being designed, it must be kept in mind that the (theoretical) live loads on the upper floors of the building do not counteract the buoyancy of the displaced water. The basic weight of the building alone may not be enough, at least in theory, to hold the building in place.

There is a special hazard in foundation work that must be investigated carefully before any decision is made on how to shore or protect the excavation. Some of the work, such as opening up of the site in a deep excavation, can disturb the local water table, which in turn can cause serious settlement and structural damage to adjoining structures, which so far have been in a state of hydraulic equilibrium. Dewatering an excavation could cause such a condition because water would be drawn from under these areas and would cause compaction of the soil in the voids left by the water drawn away. Spread footings would settle and there would be corresponding structural damage. Appropriate measures must be taken before starting any work that could possibly upset the water table. It is possible for damage to occur at quite some distance from the actual construction site. It all depends on the subsoil formation in the area. It may, for example, mean using steel sheet piling to isolate a site when other shoring methods might seem to suffice. Similarly, tube piles and caissons may sometimes act as wellheads and may create problems in terms of the water table as well as for the piles or caissons themselves.

UNDERPINNING

When the foundation of a building, other than piles, reaches a lower level than the foundation of the adjoining structure, the latter must be underpinned to prevent it from either failing or sliding into the excavated area. This may also be necessary when the new foundation work causes a shift in the water table, eroding the seat of adjoining footings, which could cause them to settle and seriously damage the existing structure.

Under certain conditions, instead of underpinning, the contractor may try to contain the soil under the adjoining foundation by using conventional shoring methods, such as steel piles and wood lagging, but this is not always possible, and depending on site conditions, it may not be the proper or acceptable solution. This is the case especially when the adjoining structure is a multi-story building or otherwise heavy structure.

There are various methods of underpinning. The choice depends on a number of factors, including the following:

A. Difference in elevation between foundations in existing structure and new structure
B. Nature of geologic subsurface conditions
C. Water table and drainage pattern
D. Contractor's available equipment
E. Miscellaneous economic and technical considerations

When all of these are summarized and analyzed, they will probably result in what the contractor considers the most economical or speediest, or both, approach.

One of the most common methods is to divide the foundation to be underpinned into a number of sections. Staggered sections are then excavated either to rock, if any, or to the level of the new foundation underneath the existing footings. Care must be taken to leave the section narrow enough so that the footing spanning the gap acts as a temporary beam and does not collapse and take the overlying structure with it. The resulting gap can now be filled with either plain or reinforced concrete and will act as a vertical extension of the existing footings. Adjoining sections are treated the same way until the whole foundation is supported.

Under certain conditions, instead of filling in every gap, some of the gaps can be spanned by what becomes a new reinforced concrete beam under the existing foundation, and only staggered sections underneath that beam reach the desired level at the bottom and become effectively extended columns under the foundation (Figure 3–7).

An alternative method that may be applied under special conditions is to use piles to support the foundation. The piles then act as a beam. These special piles are usually jacked into place hydraulically, since they obviously cannot be driven. They can only be used if they sit on solid bearing such as rock and if their length is limited; otherwise, they may lack the necessary stability or strength in bending.

Underpinning is both costly and time-consuming, and it tends to delay the general excavation because a berm usually must be left

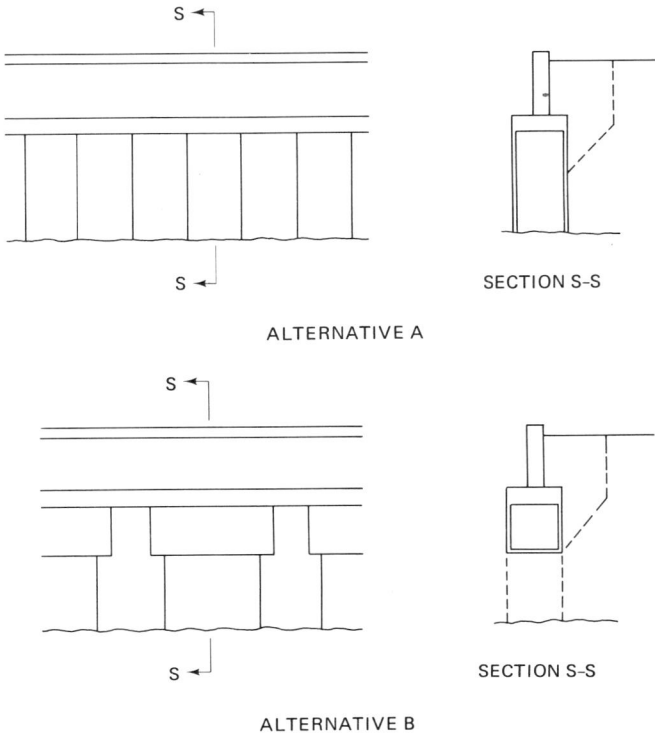

ALTERNATIVE A

SECTION S-S

ALTERNATIVE B

SECTION S-S

FIGURE 3-7. Underpinning.

against the existing building. This will be removed gradually, but since the underpinning involves a great deal of hand excavation and manual labor, generous allowance must be made for this in scheduling the work.

SHORING

When a building that has a deep basement is located on a restricted site, the sides of the excavation must be supported to prevent them from caving in. Several of the ways this can be done are discussed below.

Steel Sheet Piling

Interlocking steel sheet piling can be driven into the ground around the perimeter of the site. As excavation proceeds this sheet piling is braced with walers and diagonal bracing or tie-backs, which sometimes are drilled into rock and at times may also consist of pre-stressed tendons.

If the sheet piling is abandoned, the foundation walls can be poured against it and it becomes the exterior form — a rather expensive material to use for this purpose, but under certain conditions this cannot be avoided. In this case, the bracing has to be cut at some time during construction and the resulting holes in the wall have to be grouted in.

If the sheet piling is to be salvaged, an outer form for the foundation walls has to be built on the inside of the sheet piling. After the sheet piling has been extracted, the resulting space outside the foundation walls has to be filled with a suitable material.

Wood Sheet Piling

This is generally limited in application and is seldom used for buildings. It is more commonly used to protect trench excavations for utilities. It could conceivably be used in the case of fairly soft soil conditions and limited depth, in which case it would probably become the exterior wall form for the foundation walls.

Steel Piles and Wood Lagging

As an alternative to sheet piling, H-beam piles are driven to below the basement elevation. As excavation proceeds horizontal planks, called *lagging*, are inserted, worked into place, and anchored behind the flanges of the H-beams. The H-beams are braced by walers and diagonal struts or tie-backs. Since the struts tend to interfere with work on the site, tie-backs are usually preferred if they can be used. If tie-backs are used, great care must be taken not to damage existing utilities or other services and structures (Figure 3–8).

The lagging generally is left in place and becomes the exterior foundation wall form.

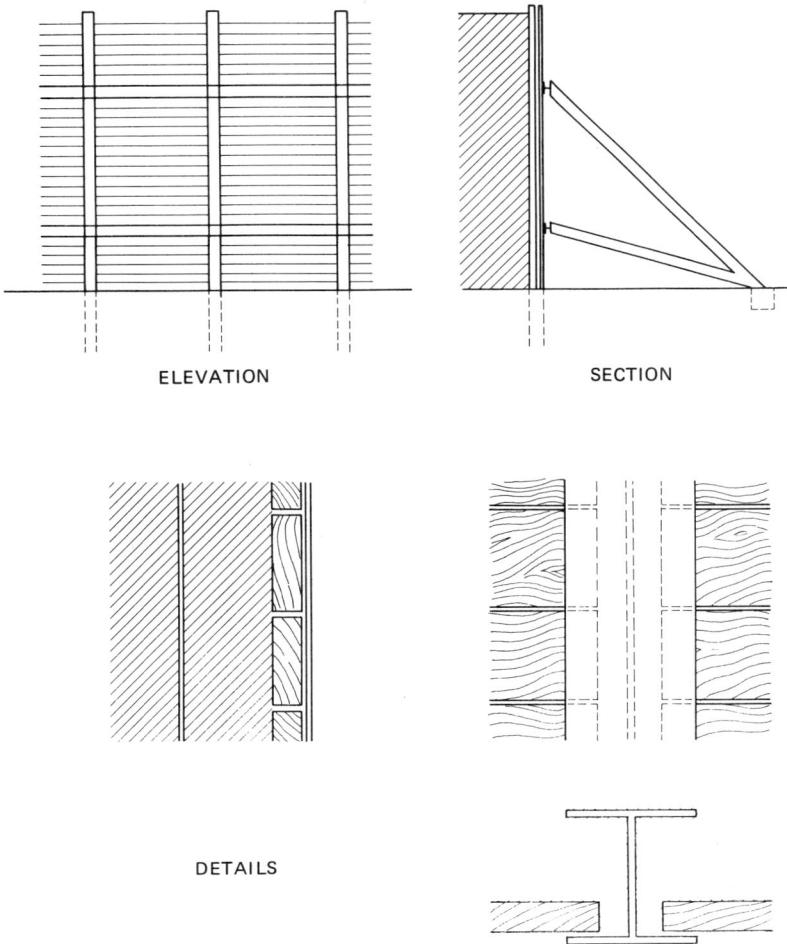

ELEVATION   SECTION

DETAILS

FIGURE 3–8. Shoring with H-piles and wood lagging.

Freezing

One method used with varying degrees of success has been to insert a series of pipes around the perimeter of the site and to circulate a cryogenic liquid in the pipes in order to freeze the adjoining bank walls during excavation. This method can be very expensive.

Cementation

Another method consists of injecting a cement grout under high pressure around the perimeter in order to create a soil cement wall around the site. This method has not been particularly successful and it depends for its success or failure on the nature and variation of the soil conditions at the site.

Pile Foundation Walls

Another method that has met with only limited acceptance consists of driving or drilling a series of piles around the site perimeter and then drilling a second series of piles between the first series of piles to create a concrete perimeter wall. The piles are tubeless concrete piles created by ramming concrete down a tube which is gradually withdrawn. As a result, these piles present a rough and uneven surface and apart from requiring lateral support after the excavation goes down, they must also be finished specially on the inside to straighten out the interior face of the foundation walls. Additional concrete may have to be poured against the surface which consists of a series of round pillars. The additional costs, apart from the bracing, have made this method somewhat uneconomical.

STRUCTURAL FRAMES

Every building has a structure. The structure may be a partial one, as in the case of wall-bearing partitions, but in most cases it is an independent frame because this gives maximum flexibility for the interior layout and it permits changes without major structural modifications.

There are three basic frames which under certain circumstances and using modern techniques may, at times, be combined:

A. Structural steel frame

B. Reinforced concrete frame

C. Laminated wood frame

Private houses do not usually have independent frames. In most cases, the wall and floor systems constitute the frame and fulfill the structural functions.

The three types of frame are compared below.

Structural Steel Frame

A structural steel frame provides a relatively light structure. The large free spans made possible permit excellent flexibility for interior layouts or for large areas uninterrupted by columns. The foundations, because they carry less weight, can be smaller and lighter. In the case of piles, fewer piles may be required in clustered pile footings. Depending on applicable codes and regulations, special fireproofing treatment may have to be applied, which could be done to blend in with the interior design and finishes. Field erection can be very fast and is possible under any except the severest winter conditions when erection crew productivity will be lower. Since a certain amount of lead time is required to permit design, detailing, and fabrication of the steelwork, it is possible to use this time to construct the foundations which will carry the steel frame. When pile foundations are required, the time to drive the piles may take longer than the time to fabricate the steel. Changes are made relatively easily and can be made at a fairly late stage.

Reinforced Concrete Frame

There are three main types of reinforced concrete frame:

A. Cast-in-place

B. Precast

C. Prestressed (pretensioned or posttensioned)

Reinforced cast-in-place concrete is very economical. It is, however, limited in the free spans attainable, and the increased weight requires a heavier foundation and relatively more piles. Its fire resistivity eliminates the need for extra fireproofing treatment. It is very costly to install under winter conditions because it has to be heated and protected during placing as well as during the extended curing period until it reaches its design strength, a process which is retarded under cold weather conditions. If concrete freezes, it loses its strength and becomes structurally useless. In this case, it must be completely removed and replaced — a very costly and time-wasting undertaking.

Precast concrete is similar to cast-in-place concrete, but since it is usually fabricated in a plant, it has the advantage of fast field erection at full design strength. Thus a precast structure is immediately ready for the next phase of construction activities. Disadvantages include the difficulty of making changes or building-in inserts after a particular member is in production. Joints and connections between members must be specially designed and usually require special hardware.

Prestressed concrete is either pretensioned or posttensioned. Either can be prepared in a plant, in which case it would be a precast prestressed structure, or installed in the field.

In pretensioning, the reinforcing steel, which consists of special tendons that are high-tensile wires stranded to form cables, is placed in the forms and stressed. Then the concrete is poured around the stressed tendons and is allowed to bond with them while attaining full design strength (Figure 3-9).

In posttensioning, the concrete is poured first with special ducts in it. Then it is allowed to reach design strength. After the concrete is cured, the tendons which are fed through the ducts are stressed inside the ducts without bonding to the concrete and are anchored to the ends of the members by means of special anchors (Figure 3-10).

Prestressed members are a little lighter since they can often be made thinner and they can also reach greater spans. Once they are made, they cannot be modified, and any change requires a new member—an expensive and lengthy procedure. Shop-fabricated field erection can be done quickly and fairly independently of weather conditions, since no extra heating may be required.

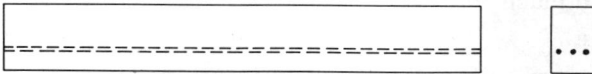

FIGURE 3-9. Prestressing. Cables are stressed and concrete is poured around them and is permitted to set while bonding to the strands.

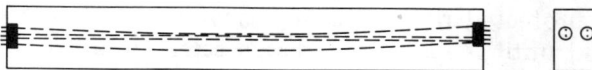

FIGURE 3-10. Posttensioning. A beam is poured with empty conduit. After the concrete has reached design strength, the tendons are stressed and anchored and the anchorages are grouted in.

Some specialized methods combine steel and concrete. An example of such a method, the lift-slab method, is described in another chapter of this book.

**Structural Steel Frame and Reinforced Concrete
Frame Compared**

In building construction, choosing between using structural steel or reinforced concrete for the building frame is an important issue. Unless certain conditions call for a specific choice, it may be advisable to consider the various factors that affect the economics of either system. Some of these factors are discussed below.

Structural steel is lighter than reinforced concrete for equivalent conditions. This means that footings and foundations can be smaller and lighter and that fewer piles may be required. Therefore, there is a savings in cost and construction time. Structural steel gives more uninterrupted floor space. Reinforced concrete is more restricted in this respect. This makes steel more economical when clear space is important. Structural steel can be erected easier and faster and it is almost independent of weather conditions. By contrast, concrete must be protected and heated in cold weather. It also takes much longer to construct because of the time required for forming, setting reinforcing steel, and placing concrete. Thus reinforced concrete results in greater job duration and corresponding higher field overhead. Similarly, the time differential may also mean less rental revenue since the building will be completed at a later date and occupancy will start later. Additional job duration and weather conditions mean higher costs for reinforced concrete. Since structural steel can be shop-fabricated while the foundation is being built, additional costs and time are saved. In reinforced concrete the structure must follow the foundations in sequence. The shorter erection time of structural steel results also in less temporary financing and, as before, in potentially increased rental revenue, since rental areas are completed sooner. Because structural steel columns are smaller than reinforced concrete columns, they require less area and result in more usable space. But structural steel may require extra fireproofing for columns and beams, resulting in additional costs. Reinforced concrete is generally considered fireproof. In fact, structural steel is often fireproofed by cladding it with masonry or concrete, either of which is expensive. Reinforced concrete requires more and additional hoisting and placing equipment, but structural steel uses hoisting for erection only; thus the reinforced concrete costs more. If cellular steel decks are used with a structural steel frame, they provide built-in raceways for electrical and telephone wiring and cables, whereas reinforced concrete floors require separate underfloor ducts. These, in turn, require thicker floor slabs, resulting in additional concrete and, consequently, in additional costs.

Generally speaking, a multi-story office building having light floor loads is likely to have a structural steel frame, but a manufacturing building having fairly heavy floor loads and not requiring any special interior finishes is likely to have a reinforced concrete frame. In the future, designers will have to make detailed computerized studies and analyses of the buildings they plan to erect to determine whether it will be better to use structural steel or reinforced concrete.

Laminated Wood Frame

A laminated wood frame is not used very often. It is more common in areas and regions that have abundant high-strength structural lumber available such as the western regions. Although the frames are lighter than concrete, they tend to be bulky and may present architectural problems. Climate and geography may impose limitations; for example, the lumber may be subject to being attacked by certain insects indigenous to a specific region.

The choice of structure depends on a number of factors and considerations, for example, the following:

- Economy
- Availability of material
- Loading to be accommodated
- Free spans required
- Fireproofing regulations
- Architectural design considerations
- Field installation limitations

LIFT SLABS

Not too long ago contractors joked about putting up a building roof first. For a number of years now this has actually become a respectable method of construction, although its application is limited.

The method consists of combining both reinforced concrete and structural steel as the structural elements of a building (Figure 3–11).

The ground floor is poured and completed first. The upper floors, ending with the roof slab, are then poured in succession on top of each other, with each successive slab becoming the form for the next

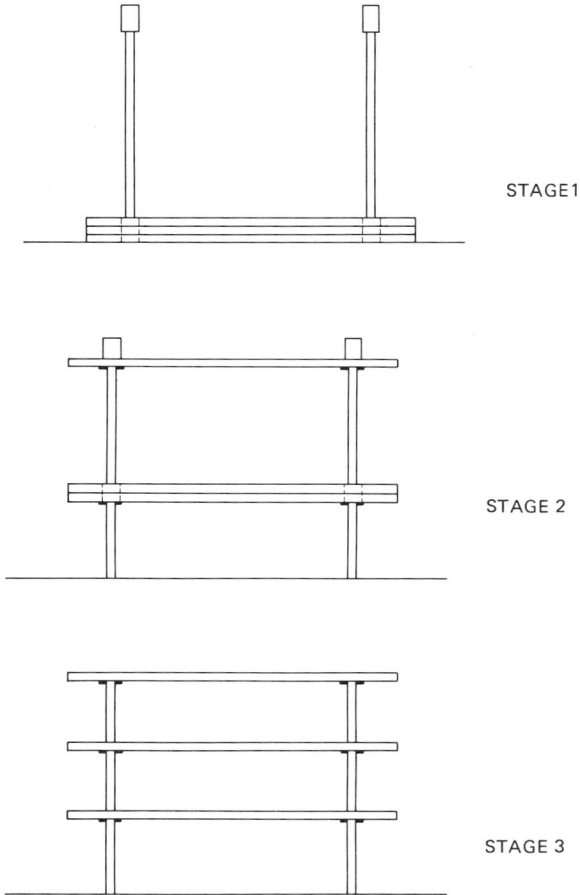

FIGURE 3-11. Lift Slab Sequence. (a) Stage 1: Slabs have been poured. Jacks sit on top of columns. (b) Stage 2 (intermediate): Top slab is installed and anchored. Second slab is parked on first slab which is anchored. (c) Stage 3: (final): All slabs are in place and anchored. Jacks are removed.

one. Suitable materials must be used to separate the slabs and prevent them from bonding to each other, and sufficient time must be allowed between pours to let the underlying slab reach a minimum strength. Box-outs are left for the steel columns, which have a uniform section for their full height. After completion and curing of the slabs, specifically the roof slab, hydraulic jacks, sitting on top of the columns, are used to jack up the slabs to the right elevation where they are permanently anchored in place to the columns by means of special brackets and anchors, after which the box-outs are filled in with concrete.

The height and number of slabs that can be erected like this are limited. The building may also require additional sway bracing or shear walls to ensure stability or to take care of wind or earthquake loads.

When several slabs are lifted, they are sometimes nested at an intermediate level. The roof and upper slabs are lifted to their final elevations.

The lift-slab method may be combined with prestressing the concrete to save on weight. In this case, either pre- or posttensioning techniques can be applied, since the slabs must reach full design strength before they are lifted. Generally, a higher strength concrete is used in lift slabs and the columns also are intentionally overdesigned to include a safety factor during construction.

The method effects a saving in formwork and also in construction time, but this saving is offset by other disadvantages. Since no beams can be used, the slab must be a true flat slab design and will, therefore, tend to be thicker, and, hence, heavier than a slab used in beam-and-slab design. Irregular recesses or projections on some floors may be harder to form, particularly in the case of protruding sections on upper floors.

After the slabs are anchored in place, construction continues along conventional lines.

TILT-UP WALLS

This is a special method of building a section of concrete wall without using forms, except for the end strip forms. This method can only be used under certain conditions. First, there must be an existing perfectly flat floor serving as the base. The wall reinforcing and any frames or openings required are then placed in position and the concrete is poured directly over a layer of material which prevents the new concrete from bonding with the existing floor. The concrete is permitted to set and cure. Once the concrete has reached full strength, the wall is tilted up into a vertical position and secured in place by means of special anchorages which afterward may be grouted in with special high-early strength grout. Curing of the slabs and lifting are sometimes done by using special vacuum equipment. Generally, several tilt-up walls at right angles to each other may be used for mutual support, but it must be kept in mind that they may have to be poured in sequence and that temporary shoring of some units may be needed until the others are ready to lend support (see Figure 3–12).

FORMWORK

Concrete formwork is one of the most important items of temporary construction equipment simply because it is concerned with our most common building material — concrete. The nature of the use of forms has resulted in an ongoing search for improving their construction, materials, connection methods, and handling. Because of the diversity of concrete elements in a building and the innumerable variations in their physical dimensions, this search has had to confine itself to economical methods to the greatest number of elements and with maximum usage to warrant their initial investment.

There are primarily two types of prefabricated forms: (A) flat panels for flat horizontal or vertical surfaces and (B) round column forms. All other formwork, as well as fillers, must still be custom-built in the conventional manner.

Flat panels are either made of plywood, all-steel construction, or a combination of steel frames having plywood facing. There are different patented systems of connecting them, all of which are designed for speed of assembly and dismantling. Some of these systems are not as effective because they are affected by the additional pressure of the concrete in a filled form, making it more difficult to strip the formwork.

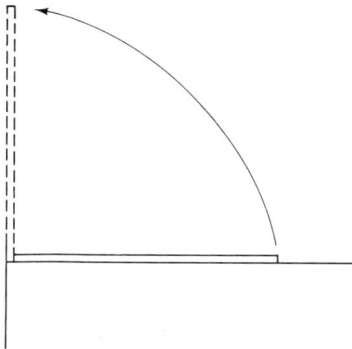

FIGURE 3–12. Tilt-up slab. Slab is poured in place and then tilted up and secured in place.

Column forms have moved from round metal forms that require a large inventory of different sizes to fiber forms that are easy to strip (some are peeled off) or that can be left in place and finished on a permanent basis.

In addition to these items, a number of specialty forms have gone through mutations in the course of time. Ribbed and waffle floors were originally built with terra cotta blocks. Later they were built with metal pans. For quite some time now they have been built with plastic pans that can be reused or with specially treated cardboard forms for single use.

Various types of scaffolding, jacks, extensible trusses, and other supports have seen little change over the years.

Perhaps one of the most startling new formwork methods involves using air structures which are domes or other shapes made of strong fabrics or plastics that are blown up pneumatically. They serve as the basic form for shaped, reinforced, thin-shelled structures upon which the concrete is usually pneumatically applied. After the concrete has reached sufficient strength, the inner skin is deflated and can then be folded and stored as a compact package.

A few formwork specialties are described briefly in the following sections.

SLIPFORMS

Some buildings contain vertical reinforced concrete elements which lend themselves to construction using the slipform method. Although this method is more commonly used in the construction of silos, grain elevators, bridge piers, and other heavy structures, it lends itself at times to the construction of elevator and stair shafts usually found as part of the service core in high-rise office or apartment buildings.

Basic requirements for slipforming are a constant vertical cross section without projections and generally of a thickness of 6 in. (15 cm) or more and a structure high enough to warrant the cost of setting up the installation.

In slipforming, concrete is placed continuously in a form which moves up vertically at a constant rate of climb until the structure is completed. Thus the forms act as an extrusion device for the concrete wall. Reinforcing steel and metal inserts can be installed in the concrete at the same time. Openings can be framed and built in, with the frames suitably strengthened to carry the weight of the overlying concrete, and in extreme cases minor projections can be accommodated by special adaptation of the forms. Depressions can be easily formed by inserting suitably shaped blocking in the forms.

The wall forms and work platform are first built in place on the site. The forms are usually built with tongue-and-groove planks arranged vertically in order to reduce friction, with the outside form slightly higher to prevent splashing and spillage of concrete. The bottom has a slight outward camber to permit easier extrusion and separation of the concrete. The forms are usually specially treated to reduce or control the swelling of the lumber which would be caused by the moisture absorbed from the wet concrete. The forms are held in place by specially constructed yokes equipped with jacks. The yokes must be high enough to permit the setting of steel and inserts in the forms. In earlier days the jacks were manually operated screw jacks climbing on threaded jack rods sitting in the concrete. Modern jacks have gripping devices and climb on smooth jack rods. They can be operated and controlled either pneumatically or electrically, but the majority of installations use a hydraulic master control system which takes up and compensates for differential movement between jacks. The jack rods can either be left in the concrete or recovered and reused by having them pass through a piece of pipe fastened to the yoke. Creating a slight space around the jack rod prevents it from bonding with the concrete, thus enabling extraction of the jack rod (Figure 3-13).

The forms must be stiffened with wales on both sides to counteract the pressure of the concrete. The work platform must be designed so that it is strong enough to handle the differential loads from concrete buggies and so forth. After the platform has moved up a few feet, a walkway scaffold is hung from the forms for the cement finishers. This may be on either side depending on requirements.

Concrete is usually supplied either by crane or hoist mounted with a suitable hopper and bucket.

The capacity of the jacks dictates the number of jacks required to carry all the loads. Consequently, this also determines the distance between them which is generally kept to under 10 ft (300 cm). The rate of climb for average projects varies between 6 in. and 12 in. (15 cm and 30 cm) per hour, but by using high-early strength cement, 18 in. (45 cm) and more can be achieved although this is generally more applicable to heavy construction projects. The forms may be from 3 ft to 6 ft (90 cm to 180 cm) in height with 4 ft (120 cm) used on average. The height is in direct proportion to the rate of climb which must be designed to permit the concrete to reach sufficient strength to support the superimposed loads resulting from the slipform operations.

Tolerances may be determined from applicable codes and are usu-

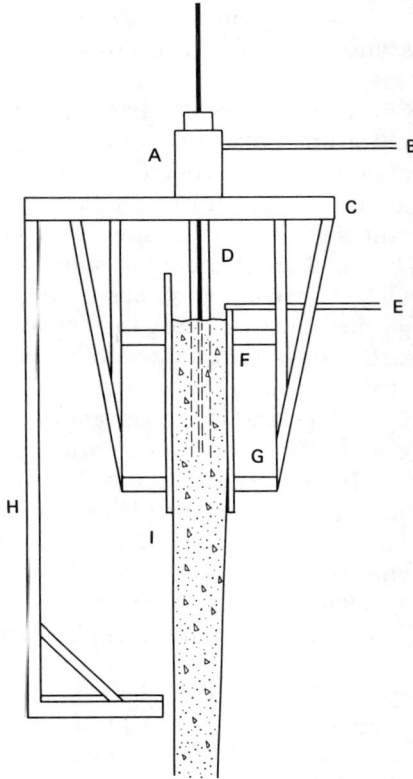

A: JACK
B: HYDRAULIC CONTROL LINE
C: YOKE
D: JACK ROD (IN PROTECTIVE PIPE)
E: WORK PLATFORM
F: FORM
G: WALES
H: FINISHERS' SCAFFOLD
I: CONCRETE WALL

FIGURE 3-13. Slipform construction.

ally derived from professional standards. They average around 6% for wall thickness and 2% for deviation from the plumb line.

Since the operation can continue right through the night, lighting must be provided in all work areas. In order to keep the concrete moist, fog sprays are sometimes mounted on the finisher's scaffold.

Once the pour is completed, the work platform may become the form for the roof slab, in which case it must be secured to the top of the structure with pins so that the yokes, jacks, and so forth can be removed and the holes in the platform can be closed up.

A specialized slipform application (in heavy construction) is used for the construction of high, tapering chimneys or towers. The forms are specially constructed so that they can be gradually reduced in size as they move up.

CLIMBING FORMS

A simpler method of building vertical structures without using slip-forms is to use climbing forms. This method consists of using and re-using a set of forms in vertical sequence, thus permitting a series of successive concrete pours. Depending on the design, the cold joints between them may require extra reinforcing steel. The forms are designed to be raised by a crane as units and placed into the next higher position being supported from the previously poured section. This method permits operating during regular working hours and avoiding the higher labor costs for shift, night, and weekend work and it also permits interruption for bad weather conditions.

FLYING FORMS

In high-rise buildings having a flat slab design and uniform floor heights, flying forms are often used to form the slabs. These forms are prefabricated sections of deck forms mounted on connected scaffold frames which after being used on one floor are lifted to another floor for reuse as a unit. So that they can be removed easily, they are equipped with short jacking devices, often in the form of screw jack legs. In order to leave the form in place for the required time, it may be found expedient to use two sets of flying forms and leapfrog alternate floors. After the flying forms are removed, ordinary jacks are used for reshoring the slabs according to requirements.

UNCONVENTIONAL BUILDINGS

There are a number of somewhat less conventional methods of putting up buildings which because of their nature are primarily engineering projects and technically belong in heavy rather than building construction. Some of these are designed for visual effect rather than for economy and probably pay a premium for this effect. Some of these methods involve primarily the roof structure whereas others involve the whole building.

Special types of roof include thin-shelled domes (Figures 3–14 and 3–15), folded plate roofs (Figures 3–16 and 3–17), cable-suspended parabolic shapes, and cantilevered structures all aiming for maximum uninterrupted floor space. Sports complexes, in particular, are designed so that there are no interior columns. Some more complex designs have retractable roofs which when closed create their own internal climate conditions.

Among the more unusual design concepts are buildings whose floors are fastened to and suspended from a central specially reinforced core which carries the weight of a number of stories hung from the top floor down. In raising the floors they may be lifted off the ground, after prefabrication, by balancing them by means of counterweights during lifting operations, similar to a lift-slab project, except that here a lot more is lifted. This means that the central core must be designed for a multiple floor load stress during construction.

Other projects unofficially being designated as habitat types, which take their name from the project originally built in Montreal in 1967, consist of individual housing modules stacked on top of each other in pyramid fashion so that each unit preserves a certain individuality and privacy in terms of its surroundings and yet shares common services and facilities (Figure 3–18).

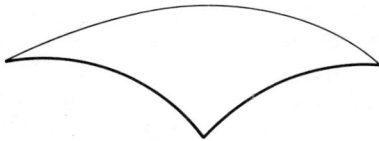

FIGURE 3-14. Paraboloid thin shell.

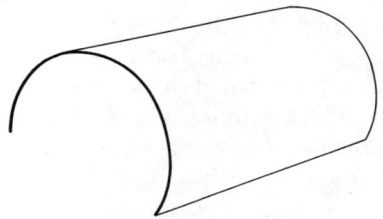

FIGURE 3-15. Cylindrical thin shell.

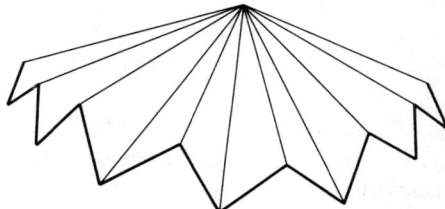

FIGURE 3-16. Circular folded plate.

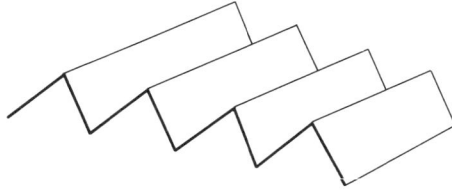

FIGURE 3-17. Linear folded plate.

FIGURE 3-18. Habitat 67. Self-contained apartment units stacked on top of each other in pyramid fashion; roofs of lower units become gardens and terraces of upper units. All are arranged to give privacy.

Under unusual structures could be mentioned air structures even though these do not involve any construction at all. They consist of a dome-shaped shell made of a soft flexible material which is kept inflated by means of an air compressor which keeps the interior slightly above the exterior atmospheric pressure. This prevents the air structure from collapsing. The entrance usually has an air lock to prevent excessive air leakage and loss of air pressure. These structures are generally used for temporary purposes. In construction there is another application for them. An air structure bubble is sometimes used as a temporary form for a thin-shell dome. This dome is constructed by applying a nonstick finish to the air structure skin, placing reinforcing mesh over the dome, and then applying high-strength concrete placed pneumatically. After the concrete has reached its required strength, the air structure is deflated and removed from the dome for reuse elsewhere.

An unusual variation of air structures is a roof built of stainless steel sheets which are anchored around the perimeter of the oval-shaped or circular building. The stainless steel membrane is supported by an air cushion created by raising the air pressure inside the dome by a fraction of an atmosphere. This is done by using larger fans to inflate the pressure.

Other unusual buildings have roofs supported on catenary cables suspended from fixed towers and resulting in paraboloid roof structures.

If these unusual structures are erected in the snow belt, they face the possibility of uneven snow loads. Although most of them may be startling in appearance, few can lay claim to any real economic advantages. Being one-of-a-kind often involves special cost factors. The fact is that conventionality in buildings has generally gone hand-in-hand with economy and unconventionality may have to pay a penalty for its special status.

SITE WORK

Unless a building is completely surrounded by other buildings, such as in downtown locations, a building occupies only part of the site and requires certain site work, which may include:

A. Parking lot and paved areas
B. Walks and patios
C. Landscaping and planting
D. Ornamental features and structures
E. Ancillary facilities for above

Parking Lot and Paved Areas

In most cases, parking lots are paved with asphalt, but in a few instances soil-cement may be used for economy or if the installation is a temporary one. Soil-cement may also be used to consolidate the subbase for the finished paving course. Using soil-cement as the final paving may be false economy because the saving in initial cost may be more than offset by potential repairs.

A good paving installation has two courses of asphalt, a rough base course and a finish course, which may be applied with a large time interval between courses. The base course may be installed and used as a temporary traffic surface during construction of the building and the finish course may be installed at the end of the project at which time imperfections and damage to the base course can be adjusted and repaired.

Good drainage is of utmost importance for the durability of the paving. This means that both the surface and the subsurface require good drainage. The subgrade and the granular base course must be well compacted under the asphalt. The asphalt should be applied under proper weather conditions, if possible, and compacted according to good practice.

A large parking lot requires a drainage system and proper lighting. The drainage may be either run into the house sewer or connected to the municipal system through a separate connection. In order to avoid a muddy site, it is advisable to install the drainage system for the parking lot before anything else.

Lighting may be a function of the traffic pattern and shape of the lot. The more light standards there are, the lower they can be to achieve the same illumination, but this also means more obstructions on the lot and less flexibility in the future if the parking grid requires changes. Light standards in parking lots require a variety of fixtures, including mercury arc lamps, sodium vapor lamps, metal halide lamps, or other high-intensity discharge (HID) fixtures. These are generally activated by time clocks whose settings must be gradually adjusted to the seasons and shorter or longer hours of daylight or by light-sensitive cells. Using aluminum poles reduces maintenance and usually warrants the additional cost. They should be protected either by bumper guards or by high concrete pedestals.

The parking grid is a function of the space available. For very large lots, such as at shopping centers, the grid pattern must be tied in with the orientation of the building on the site in order to prevent traffic jams at store closing time. If a supermarket has a car order pick-up, special care must be taken with the traffic flow pattern to prevent traffic jams which could be particularly severe if the car order pick-up line crossed a main traffic artery on the lot.

If the lot is too narrow to accommodate even modules of right-angle parking, then the car spaces can be angled to reduce the width of the modules. A comparison of angle parking for a 100-ft strip shows that 23 cars can be accommodated in a 67-ft width at 90°, 19 cars in 66 ft at 60°, 15 cars in 62 ft at 45°, and 11 cars in 51 ft at 30°.

Fire department regulations may demand a fire lane in a no-parking zone. This zone should be indicated by signs. Unfortunately, fire lanes are seldom kept open unless the police ticket cars on a regular basis.

It is possible to route traffic by arranging planting areas in strategic places on the lot. This also relieves the monotony of the asphalt (Figure 3–19).

Traffic grid lines are usually painted on with special traffic-resistant paint. The temperature differential under these lines sometimes causes hairline cracks to appear parallel with these lines, but they do not usually cause any problems.

Exits from parking lots must be carefully located relative to the existing road system. For most buildings, the exit may be a ramp leading into traffic but in some shopping center lots the exit may be

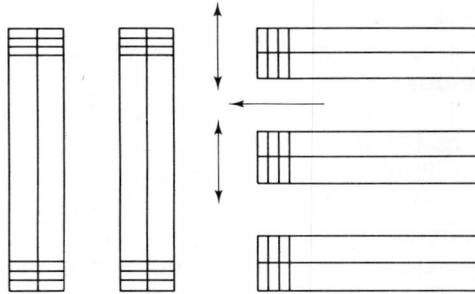

PARKING AT RIGHT ANGLES CREATES TRAFFIC PROBLEM

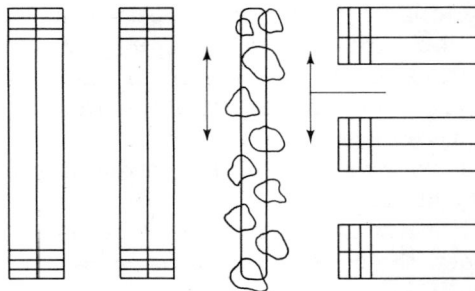

PLANTING AREA IS USED TO ELIMINATE TRAFFIC PROBLEM

FIGURE 3–19. Use of planting area to regulate traffic.

more complex. Because of traffic build-up, allowance must be made for both deceleration lanes for entering the lot and acceleration lanes for leaving the lot onto adjoining expressways. Depending on the circumstances, a traffic light may be required if there is only space for direct access into traffic.

Walks and Patios

Walks and patios may be designed to permit congregation and/or circulation of larger numbers of people and they can also be planned to direct and control pedestrian traffic. They may be provided with radiant heating for ice and snow control, but this not only can be very expensive but it also uses a substantial amount of energy. It is advisable to have a skidproof surface which sometimes can be combined with an esthetically pleasing rough-textured exposed aggregate finish. The concrete used should have air-entraining for greater durability, and proper expansion joints are a must. Wheelchair ramps should always be provided. It is usually advisable to paint them in a bright traffic color, both to indicate their location and to alert people against stumbling over them.

Landscaping and Planting

This is entirely a question of the taste of the designer. Nevertheless, there are a few guidelines to follow. The best plants to use are usually the ones that grow locally and are acclimatized and hardy. Although flowers and shrubs look very nice, they can be subject to vandalism or theft. Grass areas unless protected against being walked on can look very bad if they are used extensively for short-cuts. Protective ornamental corner fences may alleviate this problem.

If the grass area is extensive, it may be advisable to install an automatic lawn sprinkler system. This must be closed down and emptied for winter. The pipes should be flexible plastic and should be installed with sloping runs for easy draining. If they are left full of water in the winter, they will freeze and burst.

Ornamental Features and Structures

There are many ways to dress up buildings and the site around them. Fountains and waterworks have often been used in imaginative forms, but it should be kept in mind that these can be extremely costly. Water, especially in motion, has always fascinated people.

Reflecting or wading pools may enhance what could otherwise be rather drab sites, but these features require fairly complex installations.

Patios can be furnished with ornamental benches which can be both useful and ornamental. They should also have waste receptacles in sufficient quantities.

Flagpoles sometimes tend to give the building a special aura, as does artwork prominently displayed on pedestals or protected by fencing. In such a case, the artwork should be scaled to suit the size of the building and appear in character and harmony with it.

RAILWAY SIDINGS

Some industrial buildings require railway sidings and some of these sidings are sometimes inside the buildings themselves. The siding usually requires a track countersunk in the floor. The track may also be installed flush with the exterior roads and paving. It is important to ensure that the ballast under all these areas is especially well compacted so that there will not be any settlement from the heavy loads exerted by the track in use, which could result in severe damage to the floor or paving. Good drainage under the exterior track prevents frost heaves from moisture trapped underneath.

The siding may be installed by the local railway company or by contractors specializing in trackwork. In the latter case, the railway company usually takes care of the connection to its system by installing a switch and frog where the siding meets its own tracks.

WALLS AND PARTITIONS

Walls are generally the solid part of the exterior of a building, although windows technically are included, and fall into two categories: bearing and nonbearing.

Bearing walls are historically the oldest but they are not too common any more, they carry the floor or roof structure, and they are usually built of brick or concrete block. If they consist of reinforced concrete, then they are generally considered part of the concrete structure.

Nonbearing walls, technically speaking, are the different types of curtain walls. Although they are generally prefabricated panels made of precast concrete or glazed metal frames and panels made of various materials, they can also be made of brick facing with block backing or similar construction. Housing using brick-veneer frames combines brick facing with a stud or plank frame which forms the loadbearing part of the wall. The thickness and height of the wood frame are usually covered under local codes.

Partitions are generally found in the interior of the building. They, too, can be bearing or nonbearing.

Bearing partitions are rather rare and are more likely found in conjunction with bearing walls or in supporting isolated elements such as mezzanines. Certain partitions, such as those around boiler rooms and transformer vaults, may be built like bearing partitions but they do not actually carry any loads.

Wall finishes have no effect on the bearing properties of walls and partitions except to add their own weight. Exterior walls usually require additional insulation. This may be either in the form of batts or rigid insulation of mineral wool or fiberglass applied between studs or furrings, styrene materials or cork glued or fastened to masonry or concrete, or foamed-in-place materials in wall cavities, which can also be filled with loose granular insulation. Another form of insulation is sprayed-on limpet asbestos.

Nonbearing partitions are either built of masonry (brick, concrete block, glazed or unglazed tile, gypsum block, siporex, terra cotta) or built as stud partitions with wood or steel studs. Wood studs are used primarily in housing because many building codes prohibit the use of wood because of its combustible nature. Another type of partition used in commercial and industrial buildings, and sometimes also in apartment buildings, is solid plaster partitions consisting of a skeleton of metal lath plastered on both sides to form a thin but fairly strong partition. These plaster partitions are generally used in areas in which there are space restrictions.

Finish materials include plaster applied either directly on insulation or masonry or on lath which can be either gypsum board or metal lath, applied on wood or metal furrings. Gypsum board with taped joints is another possibility because it can serve as a finished surface ready to paint or as a base for plastic or other wall coverings or for tile. Plaster, too, can serve in either capacity.

Ceramic tile can be applied in several ways. It can have its own lath and rough setting coat applied directly over masonry or on studs, or it can use plaster or gypsum board as a base. The first method results in a more durable installation. If epoxy finish is used, it requires a plaster or gypsum board base which should be very smooth since the coating will tend to follow any rough contours.

The bottoms of walls and partitions require finishes at the joints with the floor. This finish can be a resilient base when a resilient floor finish is used or a wood or tile finish when wood or tile floor finishes are used. In some rare instances a special custom-made metal base may be required. It is very important to seal the space between floors and exterior walls to prevent the passage of smoke or fire in case a fire should break out. Since this joint may be hidden under a radiator or convector cabinet, it must be carefully inspected during construction.

During construction the partitions are laid out on the floor and mechanical roughing-in may be installed before the partitions are framed or the exterior walls are lined. It is expedient to find and remove any obstructions and correct any dimensional errors at this time. It is much easier to correct these problems, which could get out of hand, at this point than later when the partitions are up.

CURTAIN WALLS AND STOREFRONTS

In theory, curtain walls are exterior nonbearing walls. In actual practice, curtain walls tend to refer to exterior walls of high-rise buildings and consist of metal or precast concrete framing, glazed windows, and (insulated) spandrel panels, all combined in an integrated package.

Exterior storefronts, which could be part of a curtain wall, are usually handled separately by the glass trade. Although curtain walls in metal and glass may also be handled by the glass trade, they often have their own specialty contractors. Although in earlier days curtain walls were also made with wood framing and plywood panels, most building codes do not permit the use of wood any longer for this purpose as well as for many other aspects of building construction.

Curtain walls are made from modular designs in a factory. When they are brought to the site, they can be erected quickly and since they represent a total wall system, it means that the exterior walls can be closed in within a very short time, thus permitting other work to proceed in a closed shell which can be heated and protected more easily and less expensively than in the days of wind-blown tarpaulins. Productivity of interior work should be accordingly much higher.

If the curtain wall consists of precast panels which may be insulated sandwich panels complete with integral fixed windows, it is generally advisable to have a design featuring positive seating rather than a design having the weight of the panels supported by bolts or brackets which could fail or corrode in the course of time (Figure 3-20).

The vertical mullions of a curtain wall may be used as guide rails for window cleaning equipment and may have a special cross section for this purpose. Often the window washer's platform is equipped with special safety devices which grip the mullions in case there is an emergency.

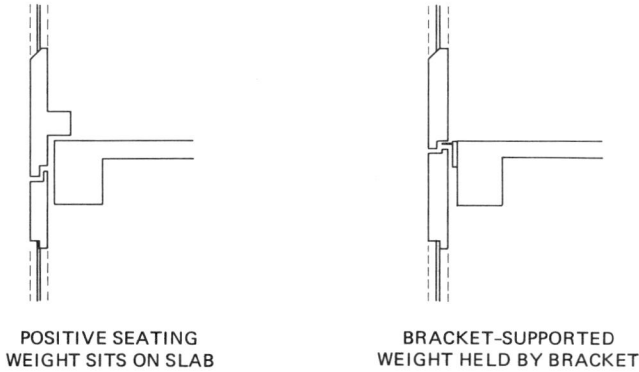

POSITIVE SEATING
WEIGHT SITS ON SLAB

BRACKET-SUPPORTED
WEIGHT HELD BY BRACKET

FIGURE 3-20. Precase concrete curtain wall panels.

Glazed curtain walls may use special tinted solar glass which becomes a design factor for the air conditioning criteria. Since this glass may have special expansion characteristics and since all glass in curtain walls is often exposed to a high temperature differential during the various cycles (day/night or summer/winter), the method of fixing these heavy panes securely and yet permitting sufficient movement for expansion or contraction can present design problems, particularly when endeavoring to keep the mullions as thin as possible. Other potential hazards, perhaps less well known, are related to the fact that every pane of glass exposed to vibrations, as induced by earthquakes, planes, heavy traffic, or other sources, will tend to break when encountering the *critical frequency*. Since the critical frequency is in inverse proportion to the physical dimensions of the pane, there is less danger of breakage as the panes get smaller. This principle was put to good use in London during World War II when large show windows were protected against the shock waves from buzz bombs exploding in the vicinity by using a prop wired tight in the middle of the show window (Figure 3-21).

At times stresses left in the glass from the manufacturing process may shatter the panes when triggered by some unknown stimulus. This may have accounted for the strange case of a major skyscraper which went through several cycles of window panes actually falling out of the building. The investigating experts were unable to come up with a reasonable explanation for this happening. Damages, which were due to the multiple window replacements and the loss of revenue because the building remained unoccupiable for a long time, ran into many millions of dollars. When it was observed that just before the panes broke the panes seemed to change color or show a

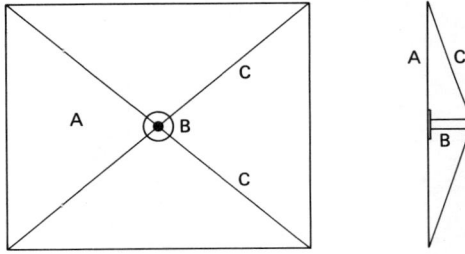

A: SHOW WINDOW
B: PROP
C: TIGHTENED WIRES

FIGURE 3-21. Method of protecting large show window against shattering.

ripple effect, the building management in a most extraordinary action arranged for special monitors to be stationed at strategic points around the outside of the building to watch the window panes and warn the security service whenever they observed a suspicious change in any of the several thousand window panes of the curtain wall. Perhaps it was an ironic coincidence that the owner of the building was a major insurance company.

FLOORS, CEILINGS, AND ROOFS

Floors, ceilings, and roofs can be considered as consisting of two distinct parts: structure and finish. Both of these must be compatible and complementary. They vary for different types of structural building frames and they are also different for floors on grade or suspended floors.

Floors on Grade

Floors on grade are either basement floors or floors in buildings that have no basements. The structural part in most instances is a concrete slab on grade, reinforced with wire mesh over a granular subbase, which may be equipped with subfloor drainage in the form of perforated farm tile drains. Since this floor is usually poured independently of the building foundation and structure, it is not usually considered part of the structural frame, except with one rare exception. If the basement is under hydraulic pressure because it is well below the water table or if the building is erected in water, it may be necessary to build the basement floor as an inverted structural slab. In this case, the joints between the floor and walls must be absolutely watertight.

The finish can either be exposed cement finish, integral or with a separate topping, or have certain types of resilient flooring. Wood flooring or some types of resilient flooring are not suitable for slabs on grade because of the moisture that through capillary action can rise through the concrete and damage or lift the flooring material. In houses a built-up or false wood floor can overcome this problem and can also provide a certain amount of insulation as well as a springier and more comfortable floor. This can be augmented by adding carpets over the built-up floor.

Floors Above Grade

The structural part of a floor above grade can be either part of a reinforced concrete frame or consist of separate members either over a concrete, steel, or wood frame. It can be of precast slabs (flat, T-beams, hollow-core slabs, prestressed slabs) on a concrete or steel frame. On steel it can be in the form of a concrete topping over steel pans or cellular steel floors or steel decks. A laminated structural wood frame would probably be combined with a mill deck or heavy plank floor.

The finish part includes cement finish, either integral or poured-in-place concrete slabs or separate toppings, which could be either over concrete or would be required over steel decks or pans. These could be finished off with resilient flooring, such as asphalt tile, rubber, linoleum, vinyl-asbestos, or pure vinyl (either in tiles or in sheet form), or with carpets, parquetry, mastic or seamless floors, tile, terrazzo, quarry tile, or epoxy and epoxy terrazzo floors. Wood floors may be finished parquetry, mill floors or barwood flooring over a plywood base, or a sanded mill deck.

In standard housing the structure may be wood joists with a wood subfloor consisting of boards or plywood and a mill floor, parquetry or resilient flooring, or carpets. In more expensive houses, steel joists may be used instead of wood joists.

Sometimes special floors have to be installed in certain locations in buildings to provide for specific requirements. In some laboratories, for example, special conductive floors having anti-static properties, and which are electrically grounded, are installed to reduce explosion hazards. Computer installations require special pedestal floors that provide an underfloor space for wiring and ducts. These pedestal floors consist of elevated frames that have removable panels to provide access at all locations. The floors are also designed to accommodate the floor loads of the computer equipment.

Roofs

The structural part of roofs is similar to the structural part of floors above grade except that roofs usually are designed for lighter loads. Loads on roofs vary with geographic locations since the main variant is the snow load which has to be supported. Steel decks may not have a concrete topping and the roof insulation may be applied directly to the deck. In addition to previously mentioned structures, there are also asbestos decks that are specially designed for roofs.

If the roof is not a dead level deck, it may have a false built-up roof structure to provide slopes to the roof drains. These slopes may be built up with a wooden falsework, but since many building codes ban falsework, it is more customary to use lightweight concrete to build up the slopes.

Most buildings use flat roofs, but houses come with a number of roof types, some of which are shown in Figures 3-22 and 3-23. In

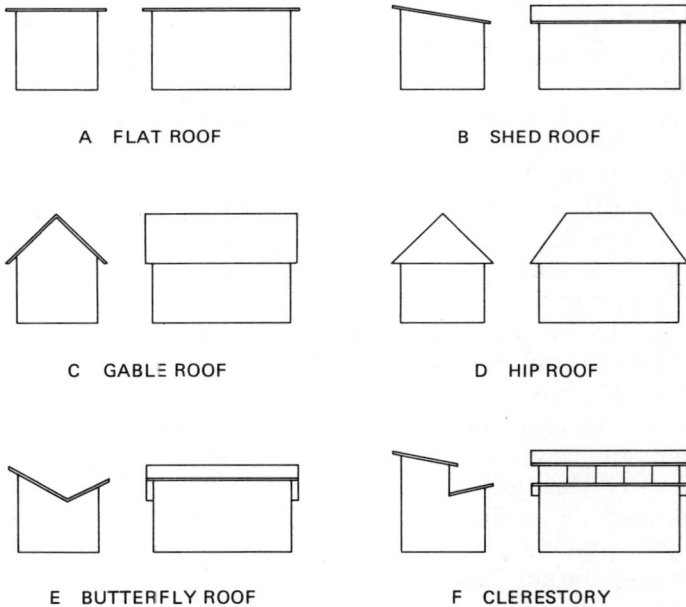

A FLAT ROOF   B SHED ROOF

C GABLE ROOF   D HIP ROOF

E BUTTERFLY ROOF   F CLERESTORY

FIGURE 3-22. Roof Types.

A GAMBREL ROOF   B MANSARD ROOF

C GOTHIC ROOF

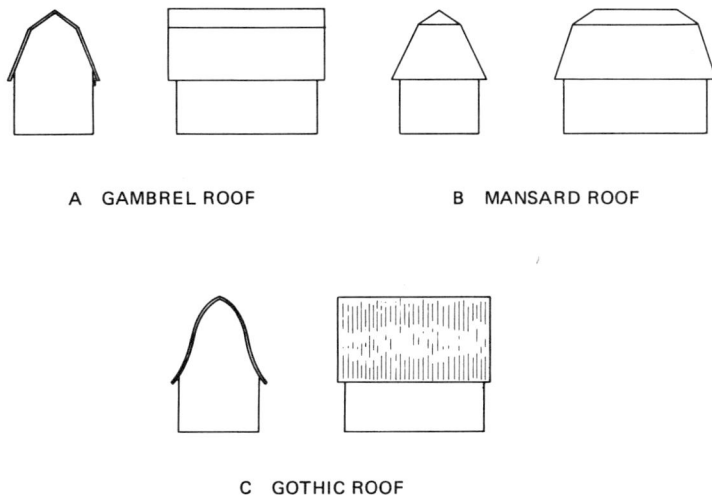

FIGURE 3-23. Roof Types.

addition, there has been a tendency toward using prefabricated roof trusses in housing. In addition to reducing field labor, they give more interior flexibility in layout, for they reduce the dependence on bearing partitions for interior layouts. Various roof trusses are shown in Figure 3-24.

There are many different types of roofs, which can be classified under two main groups:

Roofs which consist of an installation which acts as a single (large) unit, and roofs which consist of multiple units.

The first category includes such roofs as tar and gravel, and asphalt and gravel roofs which are also called built-up roofs. They consist of multiple layers of roofing felt or paper, placed and mopped in tar or asphalt and finished off with a covering of gravel. The conventional built-up roofs may contain a vapor barrier placed over the roof deck, followed by suitable insulation on top of which the roofing is built up. The deck may be concrete, steel, asbestos, or wood. Insulation may be fiberboard, fiberglass, styrene, foamglass, or other foam-type insulation material.

A less conventional roof is an inverted built-up roof. The built-up roof is applied to the roof deck and the insulation is placed on top of the roofing followed by a protective covering which can be a weatherproof hardboard, pavers, or other suitable material (Figure 3-25). With this type of roof only water-resistant insulation can be used and

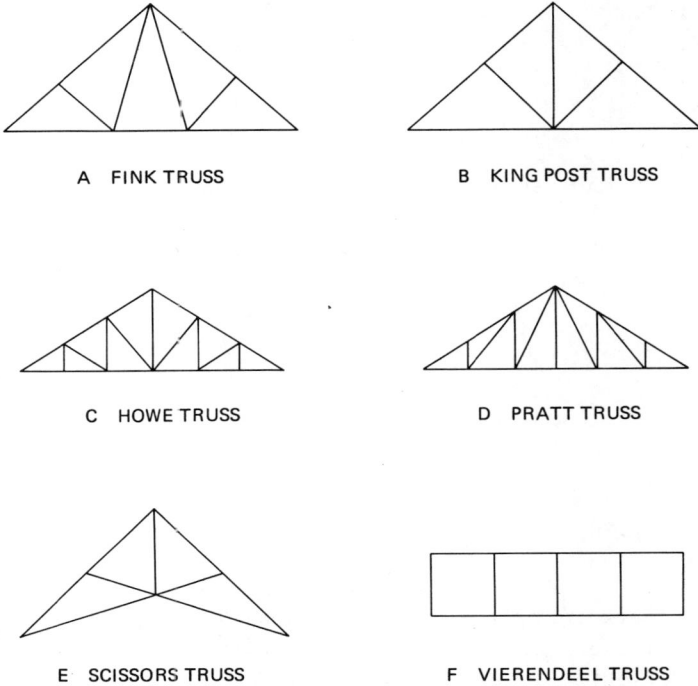

A   FINK TRUSS

B   KING POST TRUSS

C   HOWE TRUSS

D   PRATT TRUSS

E   SCISSORS TRUSS

F   VIERENDEEL TRUSS

FIGURE 3-24. Trusses.

gravel must consist of round pebble-type stones to avoid damaging the surface when walked on.

Because built-up roofs have laps at each layer, it is important to ensure that the specified number of plies are installed by the roofer so that he does not call the laps one of the plies (e.g., a 5-ply roof). This can be checked out by ordering roof cut tests, in which the testing laboratory cuts a small square of roof to count the number of ply and this patch is afterwards repaired by the roofer. This should be included in the specification.

Epoxy, plastic, or special roofing materials such as hypalon may be applied over concrete, steep, or irregularly shaped roofs. These roofing materials cannot generally be applied over insulation which must be installed underneath the roof structure. Some buildings have sheet metal roofs with or without raised seams. Sheet metal roofs are often found on steep roofs, domes, dormers, and gables on institutional or religious buildings. Roofing made of copper sheeting oxidizes over time and acquires a green patina.

CONVENTIONAL BUILT-UP ROOF

INVERTED BUILT-UP ROOF

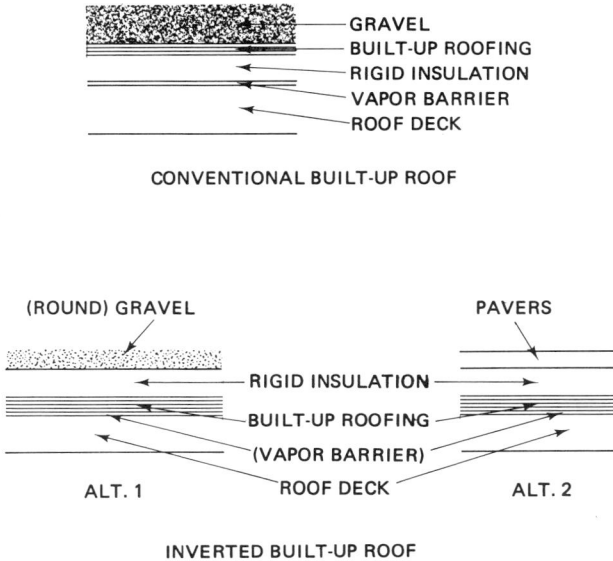

FIGURE 3-25. Built-up roofs.

All roofs use sheet metal in some form or other for flashings, copings, gutters, and downspouts. Galvanized sheet metal is most commonly used, but aluminum and copper, which are more expensive, last a lot longer.

Roofs consisting of many separate elements include shingle, slate, and tile roofs. They are often applied over wooden subroofs over heavy roofing paper. They, too, require that insulation be installed underneath. In houses, insulation in the form of batts may be stapled between the joists or rafters or installed over the attic ceiling if the attic space is not used. In the latter case, it is imperative to vent the attic space to the outside to prevent condensation and other problems.

Ceilings

In suspended floor or roof structure systems the ceilings used are either fastened directly to the surface above or have separate suspension systems.

Directly applied ceilings can be either plaster or acoustic tile on concrete slabs. On wood joists, plaster requires furrings and fiberboard or metal lath, but acoustic tile requires only furrings. Just about all other structures require suspended ceilings. These ceilings can consist of lath and plaster on a metal suspension. Most building codes ban the use of wood except possibly in housing.

Acoustic tile suspensions can be with surface-mounted or inter-locking tiles, or, in the most commonly used applications, with lay-in tiles in T-bar suspensions. The advantage of these is that the ceiling is extremely accessible at any location. In addition, these ceilings are designed to have matching modular flush-mounted (fluorescent) light fixtures, diffusers, and grilles for air conditioning and ventilation. Some fixtures have special slots that act as substitutes for diffusers. Another type of suspension on a modular grid which may be in 5-ft (1.5 m) squares uses wide channels for the suspension. They are designed to have slots for either air supply or return, and they also act as guide rails for special prefabricated partitions that are designed to go with this type of ceiling. Some lay-in tiles used in special applications have extra insulation fastened on top.

In radiant ceilings, heating coils are mounted above the ceiling which consists of suspended metal lath and plaster. Radiant ceilings can create problems if the heat is too high because the plaster will crack and will be difficult to repair.

STAIRS

Stairs or steps are required whenever there is any difference in elevation between two areas other than that which can be handled by a ramp. If the elevation is not high enough to warrant an elevator, provisions should be made for a parallel ramp to accommodate wheelchairs and baby carriages. This ramp must not be too steep or else it will become a hazard, especially if it is on the exterior of the building.

In high-rise buildings, stairs are often arranged in scissors fashion in the service core. Two stairs, which are often a building code requirement, cross each other without, however, any direct connection or access between them. See Figure 3–26.

For safety reasons, stairs and steps should have treads with toe spaces. Stairs in public buildings and exterior stairs often have straight and fairly low risers which combined with wide treads could become awkward to walk on. They are sometimes known as *monumental stairs*.

As a general rule, the combined tread and riser should average about 17½ in. (44.5 cm). In housing, stairs tend to have narrower treads and higher risers to save space, but the ratio remains the same. Headroom should be measured from the nosing of the tread to ensure sufficient clearance across the whole tread.

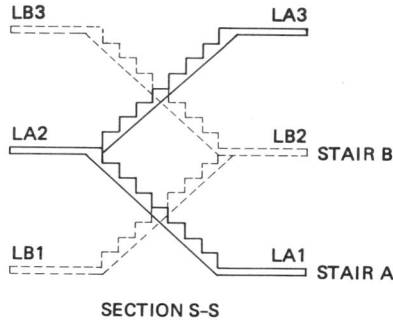

SECTION S-S

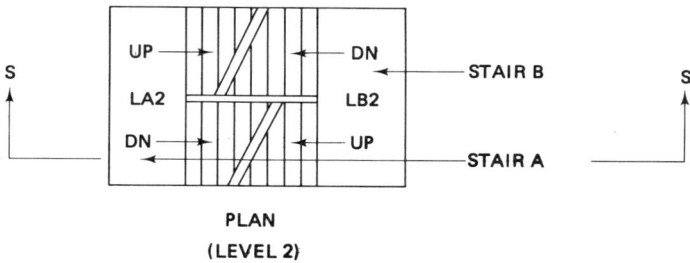

PLAN
(LEVEL 2)

FIGURE 3-26. Stair arrangement.

There are many ways to frame or finish stairs. A few are listed below.

A. Reinforced concrete: cement finish, integral or separate topping; applied terrazzo, mastic, or epoxy finish; precast treads or risers.

B. Steel: checker plate; hollow pans with concrete, mastic, or terrazzo fill; precast concrete or terrazzo treads.

C. Wood: precut wood treads and risers.

WINDOWS

Windows originally had three main functions: to provide light, air, and a view. In the course of time these functions have been somewhat modified. With modern air conditioning systems, windows often are fixed and cannot be opened, especially in high-rise buildings. This may sometimes be done for safety reasons since an open window at high altitude can be subject to strong negative pressure caused by wind currents and could endanger a person near that window.

Industrial buildings are often designed without windows because they are not required for functional purposes. It has been found, however, persons working in windowless buildings often show adverse psychological effects, and for this reason windows may be provided in plants simply to create a positive atmosphere.

In office buildings the trend is more toward full perimeter fenestration, either for esthetic reasons or to provide maximum flexibility in terms of partitioning floor space.

The aspect of light does not usually enter the picture anymore since most offices use artificial light all day long and have curtains or blinds as well.

In fact, perhaps the most important aspect these days may be the visual effect created by the windows in terms of overall building design. Nevertheless, most building codes still have minimum window area requirements in terms of floor areas, primarily for space occupied by people, such as offices, apartments, and so on. Location of windows may be especially important in residential construction where windows are also required for ventilation purposes.

Windows are made of a number of different materials, for example, wood, steel, aluminum, and plastic for frames and glass, acrylics, and other plastics for panes. The trend is more toward metal windows. Although they are more expensive than wood windows, they are also more durable. Wood windows are still used extensively in residential construction, especially in standard or lower quality housing developments. When wood windows are used, it is recommended that the best exterior grade locally available be used and that they be treated with wood preservative. An integral treatment, such as the osmose process, is generally more effective than surface applications. This treatment not only preserves the window longer, but it also reduces or prevents moisture absorption and thus prevents swelling or warping of the frame members.

Manufacturers of metal windows make special distinctions between windows used in residential, commercial, and institutional construction (Figures 3-27 through 3-30).

Commercial steel sash generally is built up with rolled T-sections in standardized modular sizes which are also designated by special codes. For example, the window shown in Figure 3-27 would be designated 65182, meaning it is 6 lights by 5 lights and has 1 opening section of 8 lights at 2 lights from the bottom.

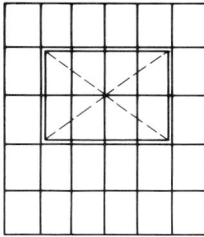

FIGURE 3-27. Pivoted.

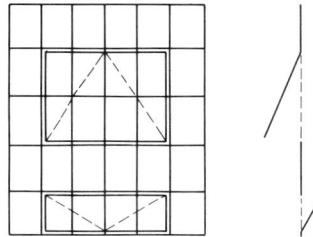

FIGURE 3-29. Hinged.

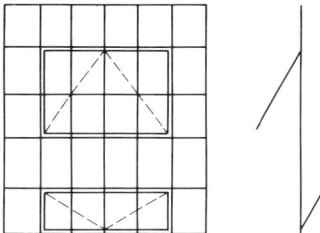

FIGURE 3-28. Security.

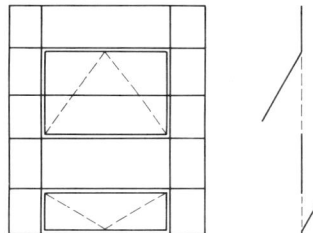

FIGURE 3-30. Architectural projected.

COMMERCIAL PROJECTED SASH

FIGURE 3-27—3-30. Window types.

Security sash (Figure 3-28) has additional bars to reduce the size of the lights and at the same time strengthen the frame members.

Residential steel windows may be built up from rolled or pressed steel sections shaped into hollow members to resemble conventional wood windows. The finish is usually baked-on enamel. This sash may come as a combination unit with screens and storm sash in a wood frame ready to install in the house.

Aluminum windows consist of extruded sections and come in a variety of finishes. Top of the line are storefronts and curtain walls which are described in another section.

Most windows have anodized finishes which keep their appearance better and longer. Some windows have provisions for thermal breaks, but they are generally found in curtain walls and storefronts rather than in residential windows.

Windows can open completely or partially or they can be fixed. Windows opening completely can be hinged at the sides (Figure 3-31) or at the top or bottom, they can pivot about their center axis, they can open vertically in sections such as double-hung windows (Figure

3-32), or they can slide horizontally as a unit or in sections. Some sliding windows consist of frameless panes only. Double-hung windows are either balanced by counterweights hidden inside the window frames or by spring-loaded spiral balances in a groove in the sash. Special types of windows are the awning type, in which a number of sections open simultaneously, and the jalousie type, in which the window consists of a number of narrow glass panes opening similarly to the awning type (Figures 3-33 and 3-34). When only part of the window opens, the vent section can either pivot about its center or be hinged. It is hinged at the top for opening out or at the bottom for opening in. Hinged vents can be combined with an insect screen, but for pivoted sash or vents, special wire cages would be required to allow for the movement out of the plane of the window. Opening sections of windows are either operated by hand or with a hook pole if they are high up, but some windows have rack and pinion or special levered cranked mechanisms. In awning and jalousie windows all

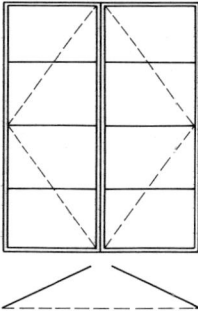

FIGURE 3-31. Casement.

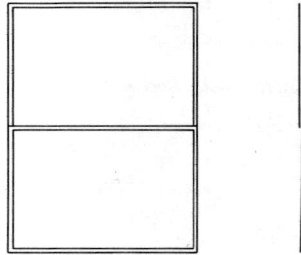

FIGURE 3-32. Double hung.

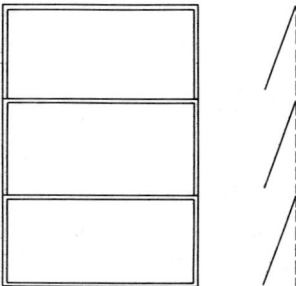

FIGURE 3-33. Awning type.

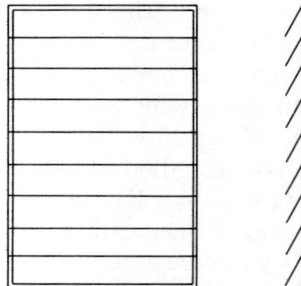

FIGURE 3-34. Jalousie.

FIGURE 3-31—3-34. Window types.

sections are joined together and cranked open or shut. In regions having colder winters storm sash is needed to provide a layer of insulating air between exterior and interior windows. Good weatherstripping is required on all interior windows, but it should also be on storm sash to help in controlling condensation. There are also other ways of providing insulation, but they have certain disadvantages. Wood windows may be fitted with sull sash which is a thin, lightly framed double window hinged to the prime window, which can be opened for cleaning. It is often used on double-hung windows in commercial buildings. Another alternative is in the glazing. Show windows and large commercial windows, as well as residential bay and picture windows, may use plate glass. To provide additional insulation, plate glass may be double-glazed and consist of two panes separated by a narrow air space. This space may be ventilated and accessible to air or it may be sealed with all air removed from the space. These units are effective, but heat or cold could be transmitted through the frame. In a show window, sealed units combined with a frame having thermal breaks is probably the most efficient combination. When only a single plate glass is used in premises in which the vapor content in the atmosphere is high, it may be necessary to install defrosting equipment that blows air across the window in order to control condensation.

Special window installations are roof monitors and skylights which are usually made of glass or acrylics. They can be either flat or dome-shaped. They usually come under the scope of the roofing trades who will install them on curbs prepared by other trades.

DOORS AND FRAMES

There is more variety in doors than in any other building component. Although doors do not represent a very large percentage of building costs, they nevertheless are a very important part since they affect circulation, traffic, and security in buildings. For this reason, they are subject to stringent codes regarding size, location, manner and direction of opening, and structure and facing materials, and fire resistance. Most doors are hinged and pushed by hand, which incidentally is mandatory in most codes for doors on emergency or exit routes, but they may be sliding, lifting, tilting, rolling, folding, revolving, bi-parting (Figures 3–35 and 3–36), self-closing, power assisted, motorized, automatic, or remote controlled. They may be constructed of wood, steel, stainless steel, aluminum, bronze, plastic, rubber, glass, fabric or combinations of any of these. They can be

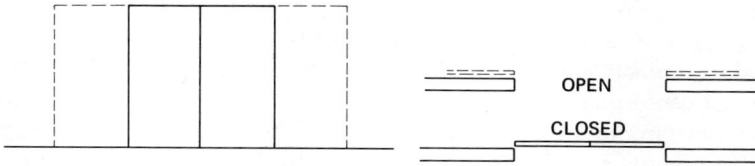

FIGURE 3-35. Bi-parting door (horizontal).

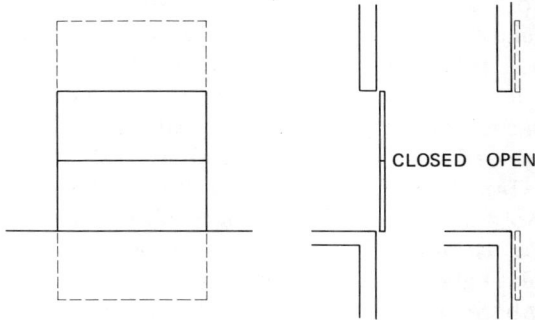

FIGURE 3-36. Bi-parting door (vertical).

solid, built-up, flush-faced, upholstered, panel, grill-type, single-piece, sectional, or slat-type. They may be individual units, with or without separate frames, or they may be part of a system such as curtain walls, storefronts (Figures 3-37 and 3-38), or prefabricated partitions.

Doors can be divided roughly into two categories: exterior and interior doors.

Exterior Doors

Entrances.

Exterior building entrances are usually designed to match the quality of the building. In high-rise buildings these doors may be part of curtain walls or storefronts, or they may be all glass, in which case armored glass is used, or they may be combination metal frames and plate glass. Depending on the quality of the building and the budget allotted to entrances, the metal used may range from anodized aluminum to stainless steel or bronze. In exceptional cases, forged or sculptured metal doors may be used.

Doors may have concealed door closers combined with special

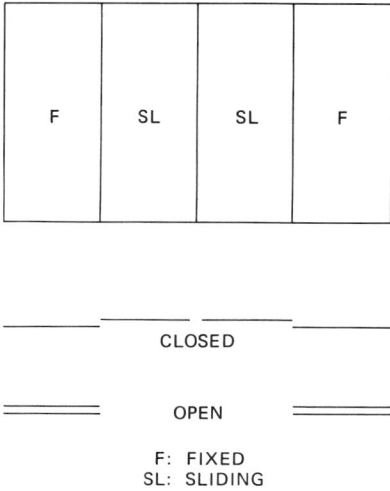

FIGURE 3-37. Storefront. (50% opening.)

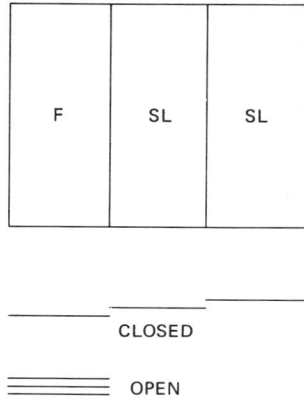

FIGURE 3-38. Storefront. (66% opening.)

pivots or offset supports. Because of their weight, they cannot be mounted with conventional hinges. They may be designed without intermediate vertical frames which can result in a finger-catching hazard and which are difficult to keep weatherproofed. The number of doors in each entrance assembly in usually determined by applicable codes and is a function of the area and/or occupancy of the building.

Doors to match the above and conform with applicable codes may also be installed when there are stores having exterior entrances on the ground-floor level.

Hardware must be suited to the particular requirements and may have to be custom-made. This could involve a serious time factor and requires an early finalization of door and frame details. The keying system must also be decided on since it affects the nature of the locks.

In lower-quality buildings, as well as in most residential construction, solid or glazed wood doors are often used for exterior entrances. These generally use conventional hinges and standard hardware.

Revolving doors may sometimes be used in entrances, but they are a dying breed. Because of their low capacity for traffic, they not only are discounted by most codes as regular exits but they may also be banned because they can be a hazard in emergencies. Revolving doors consist of four leaves which are intentionally hard to push to prevent the door from spinning too fast. Many have a safety feature that permits them to be opened for emergency evacuation by exerting a certain pressure in the reverse direction.

Man-Doors.

These serve as secondary or emergency exit doors from buildings. They are made of wood or metal, such as kalamein-type doors which consist of wooden cores covered with sheet metal, or are built-up hollow metal doors, all in matching frames and usually equipped with panic hardware. The location is often governed by applicable codes which may specify the maximum distance to exit doors from anywhere on the floor.

Vehicular Doors.

These are primarily required for shipping or receiving purposes and other vehicular traffic. They come in many types such as overhead doors, vertical lift doors, rolling doors, and other special doors.

Overhead Doors.

This is probably the most common type. Smaller and residential overhead doors are usually made of wood, either panel or flush type, and often are provided with glazed vision panels. They are usually sectional, but very small doors in private houses may be single-panel, tilt-up doors. Sectional doors may be mounted on a parallel track which makes it difficult to lift the door, but better quality doors use offset hardware so that the doors close tightly against the frames. A rubber astragal or compressible molding on the leading edge of the door ensures a weathertight fit (Figures 3-39 and 3-40).

Overhead doors for industrial plants or service stations may be made with glazed metal frames fabricated from steel or aluminum in a narrow section to provide maximum glass area for light and visibility. In special buildings such as automobile showrooms doors could be made of stainless steel.

Smaller doors are usually handled manually, but larger doors require either a geared chain hoist or a motorized operator. The latter may be key-operated such as in apartment buildings, by push buttons, or even by remote radio control from a transmitter in the car. The latter method can give a lot of trouble and is not too popular.

Frames generally match the door materials.

The door may be balanced by counterweights, a method which is

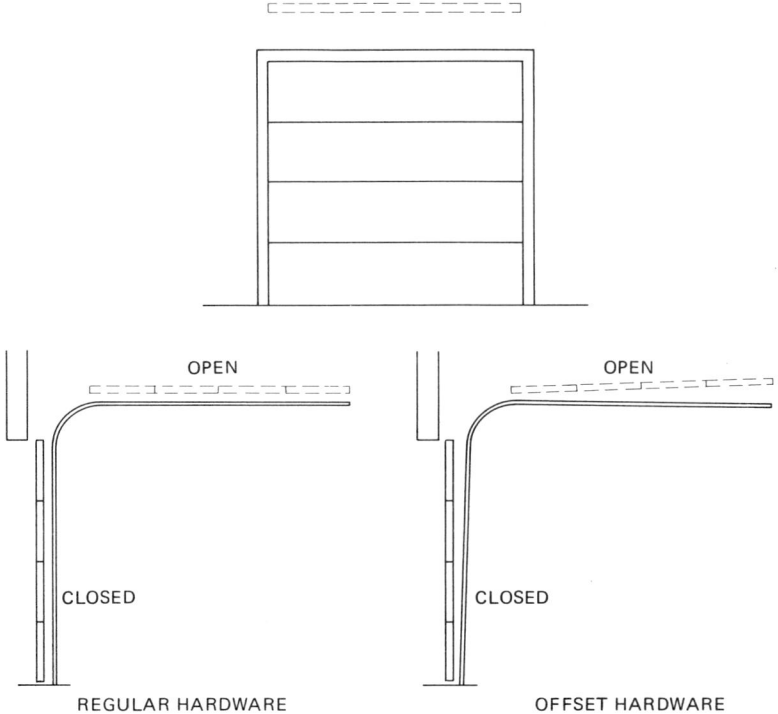

FIGURE 3-39. Overhead door (sectional).

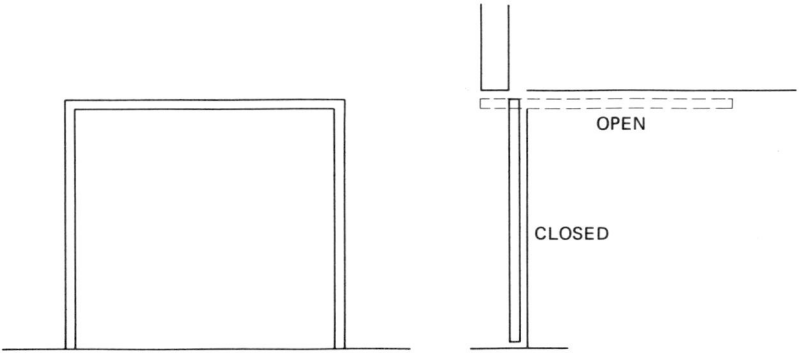

FIGURE 3-40. Overhead door (tilt-up).

somewhat obsolete, by extension springs, or by torsion springs, with the last type perhaps the most common one used.

It is important to remember that every time a solid overhead door is painted it may be necessary to adjust the tension of the springs because the weight of the additional paint may upset the equilibrium of the door requiring compensation.

Vertical Lift Doors.

Some openings that require large doors cannot accommodate regular overhead door tracks. In such cases, special vertical lift doors are required. They may consist of sections which after lifting are stacked behind each other, domino-fashion, above the opening. They can be chain operated, but they are usually motorized. They are used at times for doors over railroad tracks and in such cases must be designed to accommodate the rails (Figure 3–41).

Other types of vertical lift doors consist of rigid built-up steel panels sliding in vertical frames which may be part of the steel structure of the building, a type of door often found in heavy manufacturing mill buildings or machine shops. These doors are counterweighted and may be opened manually or, if used frequently, are motorized.

Rolling Doors.

These can be used as an alternative to regular overhead doors. They consist of wooden or metal, steel or aluminum, hinged slats which are rolled up on a drum when the door opens. The drum is inside a special housing at the head of the door. Rolling doors may be used when there are special space restrictions. They can be manual,

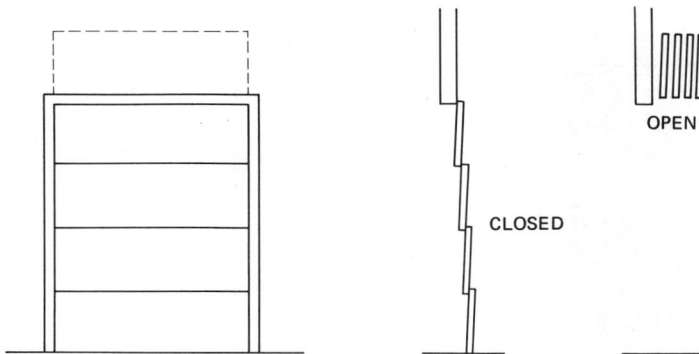

OPEN

CLOSED

FIGURE 3-41. Vertical lift door.

chain operated, or motorized, depending on the size and use. When used as a firebreak they may be equipped with a fusible link making them self-closing in case of fire. In that case, they must conform to the regulations for the fire rating required (Figure 3–42).

Special Doors.

Certain doors may be so large that they have to be custom-built and require special motorized operation. This includes doors in aircraft hangars, railway sheds, and shipyards. They are generally built-up steel doors and are designed to either fold or slide on special tracks out of the way.

Insulation.

Because of their exposure to the weather some doors may require insulation which can either be built into the door or surface applied and covered with a protective material. Generally, doors are seldom insulated since there is more heat loss, or gain, every time the door is opened in cold or hot weather.

Interior Doors

Many exterior doors can also be used for interior purposes. The main difference, apart from the finish applied, may be in the weatherproofing and hardware. Heavy ornamental doors in stainless steel, aluminum, or bronze are used primarily for visual effect. Industrial steel doors, equipped with fusible links in certain cases, are used as firebreaks, as are vertical lift doors, some metal overhead doors, and metal rolling shutters.

Sliding Storefronts.

Many shopping centers and stores in shopping arcades use sliding storefronts consisting of movable and fixed panels. In some localities

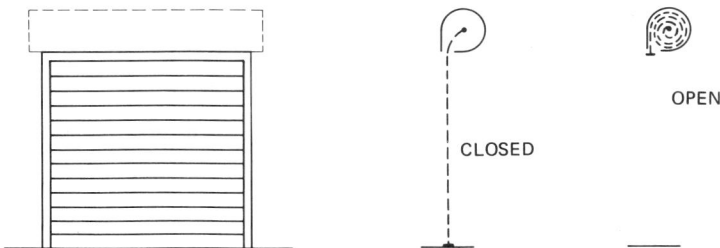

FIGURE 3–42. Rolling shutter.

regulations require a hinged emergency door in the storefront. For esthetic reasons, this door is often placed in the show window where it is fairly useless, but the regulation is being followed. Less expensive storefronts are finished in plain anodized aluminum, but more expensive design may call for colored anodized finish or chromeplating. The doors may also be arranged to slide out of sight on a sharply curved track into a special door pocket. The tracks are usually mounted under a bulkhead which becomes part of the signband. A lay-in tile ceiling provides for ready access to the sign when it needs servicing or repairs (Refer back to Figures 3-37 and 3-38).

Some stores use either vertical or horizontal grilles instead of doors, but these grilles can be dangerous because there is the possibility that lighted cigarettes might be thrown into the stores either accidentally or intentionally to damage merchandise or cause fires.

Fire Doors.

In many locations doors are required as firebreaks. These doors may be built-up hollow metal or kalamein doors, or, depending on the rating required, solid doors. They may be sliding or hinged, but all must be self-closing. Rolling metal shutters having fusible links may also be used for this purpose. Stair shafts generally require fire doors which may be provided with vision panels made with wire-mesh reinforced glass. Fire doors must not be locked in the direction of exit. If, however, the doors are intended to be locked from the other side for security reasons, they may have to be equipped with panic hardware.

Wood Doors.

Offices, apartments, institutions, hotels, motels, and houses generally use wood doors to various rooms for privacy or security. These doors are mostly flush doors. The face may consist of different materials ranging from tempered pressed boards to expensive veneers or upholstery. The core may be built up with a wood skeleton or solid which gives it a slightly higher fire-resistance rating.

Built-up doors have a wide head and bottom rail for adjusting the height of the door and a wide center rail for mounting the door handle or latchset.

Depending on the facing material doors may be painted or stained.

In some offices, doors may be upholstered or covered with fabric and even be equipped with felt weatherstripping for soundproofing.

Other more expensive doors may be paneled or sculptured. Doors in offices or institutions may have plastic facing. If the edges are to be covered with plastic laminate, then the doors must be hung first and adjusted because it is not possible to hang and adjust them after the plastic is installed.

Hardware usually includes one or one-and-a-half pairs of butts, depending on the size and weight of the door, and a latchset or lockset as required. Apartment doors may be equipped with deadlocks which cannot be opened without breaking the door.

Closet doors may have louvers to permit air circulation. Louvered doors, full height or only partial, equipped with double-acting hinges are sometimes used for kitchens to create a special effect, but they will not stop cooking odors, steam, or heat from escaping from the kitchens to other rooms. If solid double-acting swinging doors are used, for example, in restaurant kitchens, they should always have vision panels to prevent nasty accidents.

Folding Doors.

In addition to the large industrial doors which fold because of weight considerations, there are many other and considerably smaller folding doors that are used for a variety of purposes.

In offices, convention halls, and restaurants whole rooms may be divided by means of large folding partitions which often fit into specially constructed pockets.

Smaller folding doors are used when there is no room for hinged doors. In addition to the variety of finishes ranging from wood to plastic laminates to fabrics or even metals, there is a large range in quality of construction of folding doors. Inexpensive folding doors have hinges made of fabrics or other flexible materials. The best doors are equipped with pantograph mechanisms for the folding action. Closets often use bi-fold doors consisting of a pair of hinged panels.

Special Doors.

There are various doors for special purposes. These doors are discussed in the following paragraphs.

Wooden slatted rolling shutters, similar to roll-top desk covers, are sometimes used for wickets.

Flexible rubber doors are used in industrial plants. They permit carts to be pushed into them for opening them without causing any damage.

Wire mesh partitions may have matching doors and gates in steel frames.

Toilet stalls have matching doors built up of light-gauge sheet metal and finished in baked enamel.

Perhaps the strangest door of all — the air curtain — can be mentioned here although it is not a real door and it does not actually have a door. An air curtain consists of a controlled stream of air flowing vertically through a door opening which is left open all the time. Its main purposes are to prevent insects from entering and to provide a break against cold or hot air entering the premises. The speed and pressure of the moving air are such that they do not cause any discomfort to people walking through them.

HARDWARE

Although hardware represents a comparatively small cost item in a building, it is nevertheless an extremely important part of it.

Visually, as well as tactually, hardware is one of the first close-range items to catch the attention of a person entering the building and can thus trigger a subliminal impression, which will tend to create a correlation between the hardware and the quality of the building. Far more important, however, is the fact that hardware, and specifically door hardware, is one of the most used, and often abused, components of a building, since the doors are in continuous use and the wear and tear on them is far higher than on any other part of the building. Thus it is essential and economically sensible to use the best hardware available. It is equally important to choose the right hardware. Although many consultants do not like the visual aspects, doors should be equipped with door holders and door closers. Since many people do not know the difference or respective functions between a door holder and door closer, a brief explanation is given below.

The door closer is a device that prevents a door from being shut forcefully or slammed shut. A door holder, because it limits the outward travel of the door, prevents a door from being opened too far, which would impose a serious strain on the hinges as well as on the door closer. Thus a door holder actually protects both the door closer and the door. The acidity in people's skin varies greatly. It is, however, this acidity which in the course of time attacks a plated finish on hardware and wears it off. Although it is very expensive, stainless steel hardware, because of its resistance to any type of corrosion, is probably the best on a long-term basis. In addition, it is available in a variety of attractive styles and patterns.

When concealed door closers are used, the floor type is generally superior to the frame-type because of the limited size of the frame members. A heavy-duty door closer should always be used. Hinges must be chosen to suit the size and weight of the door and a sufficient number must be used. Ball-bearing hinges ease the opening of a heavy door.

Since door holders and door closers may come from different manufacturers, great care must be taken in matching them up and avoiding a conflict if they are mounted in adjoining positions.

Pressure-activated push bars that automatically open doors are very convenient for the public, especially when they are carrying things. Push bars are widely used in supermarkets and major stores, but they require a lot of maintenance and malfunctions tend to occur very often. Unless required by the nature of their location, they are not recommended.

Door handles or push bars require a design that prevents fingers or hands from being caught when opening or closing doors. The handles and push bars should be spaced away from the door at a safe distance and should not have any projecting or sharp corners or edges.

Locks should be heavy duty. Most buildings have some kind of master key system, of which there are different types. A convenient system is the "key-alike" system in which one key opens all doors, but this system may present problems with privacy, and some insurance companies may object to the system. It may be more expedient to have all public and service areas keyed alike and to have individual locks for offices, stores, and so on keyed on an independent basis, with or without a master key for these premises. In the case of a multiple door assembly, one door may have special hardware for the benefit of people who have to enter after regular hours when the building is closed. This is sometimes done for bank customers who want to deposit money in a bank night depository in a shopping mall.

Local regulations govern such items as panic hardware on exit doors from public areas.

Doors leading to areas having special ventilation systems or unventilated spaces may require louvers or vent grilles to permit air circulation.

Public washrooms are not always open to the general public. Therefore, it may be advisable to equip washroom doors with keyed latchsets.

Window hardware usually comes as an integral part of the window in the case of metal windows, and it is usually the responsibility and choice of the manufacturer. Wood windows are seldom used today except in private housing or in less expensive apartment buildings. If the windows are double-hung, the right sash balance is required to permit easy opening and closing of the windows and maintaining them at any intermediate position. Lifts and sash locks are required, and if storm sash is used, special hinge brackets, Japan buttons, and storm sash adjusters are needed. Weatherstripping and weather bars are usually built into the window by the manufacturer.

Many people are confused by the classification of doors and how to specify hardware for doors. The checklist shown below should be of help.

Checklist

DOOR HARDWARE ORIENTATION

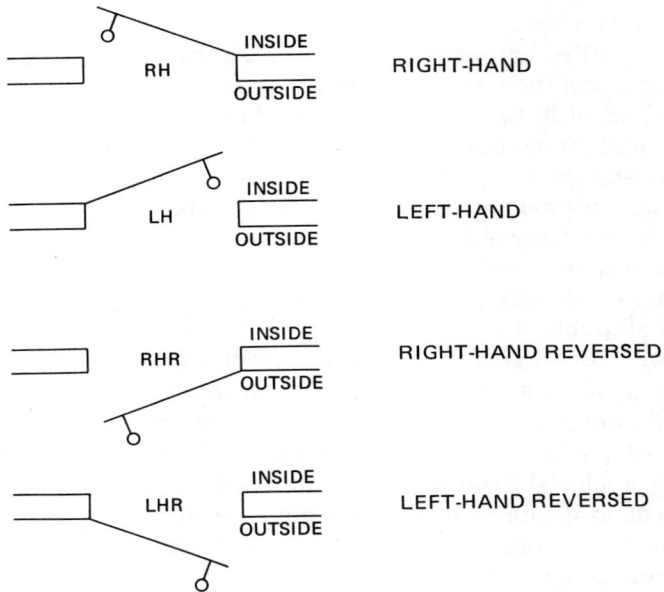

```
                       INSIDE
                      ┌──────
           RH         └──────        RIGHT-HAND
                       OUTSIDE


                       INSIDE
                      ┌──────
           LH         └──────        LEFT-HAND
                       OUTSIDE


                       INSIDE
                      ┌──────
           RHR        └──────        RIGHT-HAND REVERSED
                       OUTSIDE



                       INSIDE
                      ┌──────
           LHR        └──────        LEFT-HAND REVERSED
                       OUTSIDE
```

INSULATION

Insulation is a rather general term and as applied to buildings can be divided into three categories:

    A. Thermal insulation

    B. Soundproofing

    C. Fireproofing

It should be noted that there is a certain indirect relationship among the three categories. Certain materials used for a specific purpose of one of the above may satisfy requirements of another one. Some of these requirements may be subject to applicable codes whereas others may depend on local conditions or specific building conditions.

Thermal Insulation

Heat, which is thermal energy, can travel or be transmitted through radiation, convection, or conduction.

In radiation it moves through air or another medium until it strikes a solid surface which then absorbs a certain amount of heat.

In convection it rises in gases or liquids which are heated.

In conduction it is transmitted through various materials. In fact, all materials can conduct heat, but some do it less efficiently than others. These materials are heat resistant. Thermal insulation consists of materials which resist heat transmission primarily in one of two ways, either by reducing the conduction of heat or by reflecting radiated heat.

In earlier times many prominent buildings such as palaces and castles were built with heavy stone or masonry walls, not only for structural or defensive reasons but also to provide a certain amount of protection against cold weather. Obviously, this was not very efficient since these materials are not good insulators. The initial remedy was to place skins and rugs on the floors and hang tapestry and Gobelins from the walls, all of which served as insulation as well as decoration. Modern insulation, as we know it, originated only in this century.

Thermal insulation may be required either to keep heat in, such as protection against cold weather, or to prevent heat build-up from unwanted sources, such as steam or hot water pipes.

Insulation materials that counteract conduction usually create isolated spaces, which may be extremely small, that trap any air that is unable to circulate and act as a heat barrier. Among these materials are mineral and glass fiber products in loose granular or batt form, various styrenes, glass, and other foamed insulation, cork, diverse boards made from wood or other cellulose materials, and miscellaneous organic or inorganic substances converted into lightweight concrete products. Radiated heat is reflected by means of aluminum or other metallic foils which must be polished and shiny so that they act as a heat mirror. In order to be effective, they must have an air space on the shiny side and should not touch other materials in order to prevent heat conduction.

Insulation by itself, however, is not enough. There must also be a suitable vapor barrier. Most air, except in extremely dry regions, contains a certain amount of water vapor moving within it. Under certain conditions, when it encounters a solid surface, it condenses into water at a temperature called the *dew point*. This depends on variable conditions of temperature and the amount of water vapor in the air. The higher the temperature, the more water vapor can be absorbed in the atmosphere before saturation. At the point of saturation, a relative humidity of 100%, any excess moisture or drop in temperature causes condensation.

The thermal properties of the insulation and the wall section can move the dew point so that it falls within the insulation itself, but condensation occurs on the adjoining solid surface. Thus, in order to control this aspect, a vapor barrier must be installed on the warm side to prevent passage of water vapor from the warm side to the cold side and into the wall or roof. Since water vapor moves with air, this barrier must be airtight if it is to be effective. Conversely, the insulation or air space in a cavity wall should be vented to the outside in order to permit free passage of the water vapor when the relative humidity or temperature changes. Any condensation occurring within the wall or insulation could cause serious damage to either, and it would reduce the effectiveness of the insulation without ever being apparent. In addition to ventilating wall spaces, unused areas such as roof attics, crawl spaces under floors, and similar areas should also be ventilated in order to protect ceilings and floors.

The vapor barrier may be part of the insulation, such as in batt insulation, but it is advisable to have a separate proper vapor barrier such as polyethylene material which is immune to moisture deterioration.

Although the function of insulation is reversed in summer in any hot climate, there is usually no special problem with water vapor, because humidity is generally much higher in winter and the temperature differential in summer is only a fraction of that during winter between interior and exterior temperatures. As a result, the problem does not usually become a critical one.

Thermal insulation comes in several different types:

A. Loose or granular material used either in roof spaces or in hollow walls. It is not recommended for hollow walls since it tends to pack itself down in the course of time, resulting in an empty space on top and compacted material at the bottom, either of which does not have much insulation value.

B. Batt insulation which may be paper-backed and have a vapor barrier on the front with overlapping flaps which can be nailed between studs or joists.

C. Solid slabs which consist either of compacted fibers, or solid foam materials, or boards made from cellulose products held together by cement. They are either placed in spaces between studs or joists or they may be glued to concrete or masonry with a special mastic. In the latter case, special wire anchors may be used to ensure adhesion of the insulation if the glue is not adequate. In addition, the insulation may be coated with a layer of bituminous material. For walk-in freezers and coolers, cork is one of the best insulators, but its use for other commercial purposes is sometimes banned by local codes because when it burns it gives off lethal fumes which would create an additional hazard in case of fire.

Some slabs used for roof insulation have a rigid backing that is specially designed to receive the tar and gravel built-up roof.

D. Foamed-in-place insulation which consists of materials containing a large number of tiny air bubbles that expand under chemical action and fill the complete space. These materials are inert to moisture.

E. Foil insulation which may come with paper backing or is sometimes applied to the back of shiplapped plaster lath.

F. Sprayed-on insulation such as asbestos limpet mixed with cement, as used on steel beams and decks or concrete slabs and walls.

With both the increasing shortage and the rising costs of energy, maximum feasible insulation should be provided for buildings. In addition, exterior openings (windows and doors) must be designed to provide maximum resistance to heat loss or gain. They should be properly caulked and weatherstripped and double windows or double glazing should be used. Auxiliary aids such as window curtains also help.

Soundproofing

Soundproofing is basically insulation or isolation against sound. Since sound is mechanical energy transmitted in the form of vibrations through different materials, sound can be controlled by dampening these vibrations or changing their amplitude. This can be done in a number of ways, primarily by installing materials that absorb sound or disperse it, changing the shapes of spaces or features to avoid echos, breaking up spaces with dividers or sound baffles, or reducing vibrations by using more mass. Sound absorbent materials include resilient flooring, carpets and underlay, acoustic tiles, and various wall coverings. Partitions can be treated by covering them with sound-absorbing or sound-dispersing materials or by staggering studs in stud partitions and fastening the facing materials on alternate studs on each side. In addition, insulation batts can be installed inside the partition. As an alternative, special resilient clips can be used to fasten plaster lath or drywall (Figure 3–43). Masonry partitions can be made more soundproof by making them thicker and thus adding more mass. In addition, loose insulation or lightweight concrete materials can be added in the voids of the blocks. Sprayed-on insulation such as asbestos limpet has sound-absorbing properties and can be used on steel or concrete. It should, however, not be exposed because of its tendency to dust harmful asbestos fibers.

Sound can be pleasant or unpleasant. Unpleasant sounds would probably be considered noise, but even pleasant sounds such as music can turn into noise.

Sound is measured on a logarithmic scale. Thus in terms of acous-

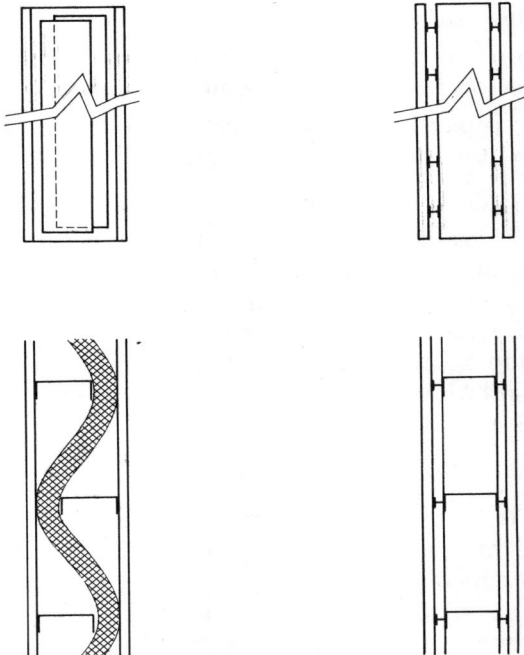

A   BATT INSULATION BETWEEN
    STAGGERED STUDS

B   GYPSUM BOARD OR PLASTER LATH
    FASTENED TO STUDS WITH RESILIENT
    CLIPS

FIGURE 3–43. Alternative ways of soundproofing stud partitions.

tic comparison, an office during business hours might have an intensity of 10^5 or a sound level of about 50 decibels as compared to a noisy plane which may have an intensity of 10^{14} or a sound level of about 140 decibels.

Fireproofing

There are two things to be considered in fireproofing: (A) the choice of fireproof or fire-resistant materials to be used in construction and (B) the treatment with special fire-resistant insulation.

Although some materials such as steel or concrete are called fireproof, at times they are really not. At best they are fire-resistant. Various building codes have special ratings for fireproof or fire-resistant construction and specifications for these classifications. Sometimes the rating is a function of the thickness of a specific material, such as asbestos board.

Steel beams and decks are sometimes insulated against fire damage by spraying on an asbestos limpet cement mixture, which can also be used on concrete. As an alternative method, steel beams and columns can be covered with masonry, or encased in concrete, or protected by using plaster or drywall covering.

Partitions in wood or steel studs may be protected with asbestos board or to a lesser degree with asbestos paper under the facing material. Masonry or concrete block partitions are considered fire-resistant and have their own ratings in proportion to their thicknesses.

Asbestos board can also be used under wooden floor systems, but it must be combined with some soft material to prevent a sound problem. In addition, any wooden members can be treated with special fire-retardant paint or chemicals.

Wood flush doors increase their ratings if they have a solid core instead of a built-up hollow core. Other types of flush doors have a hollow core filled with loose mica insulation, but this is not as effective. Some building codes give a solid core a 20-min rating which it is hoped is enough time to call the firemen in case of fire.

Effectiveness of Insulation

As mentioned above, insulation is used for various reasons. It is usually installed where its effectiveness is difficult to establish because it cannot be tested directly. With the energy situation worsening, it is becoming increasingly important to ensure that the insulation installed performs as intended. In order to test the thermal properties of insulation, a technique, originating in Europe, has been used successfully in many instances to show whether or not insulation is sufficient and has been properly applied. Thermography is based on the principle that infrared radiation is an indication of the heat emission or retention of materials. By taking special pictures, called *thermograms*, of the infrared patterns of buildings or building elements, it is possible to judge the effectiveness of the insulation. Preventive and remedial measures can then be taken accordingly. The method has found recognition to the extent that some municipal authorities insist on thermographic inspection, especially for housing projects.

X-RAY PROTECTION

In hospitals and radiology laboratories special protection should be provided to guard against radiation hazard outside the premises. The customary method is to provide lead sheets, about $\frac{1}{8}$ in. (3 mm) thick, in floors, walls, and ceilings. Doors, too, should be covered in such a manner that there is no gap left between door and frame. Conventional finishes can then be applied over the lead sheets. For partitions, the simplest method is to double them. Floors can have a topping applied over the lead. For ceilings, the lead could be installed over the suspension, but the additional weight must be allowed for since $\frac{1}{8}$ in. of lead weighs approximately 8 lb/sq ft (39 kg/m^2).

REVOLVING RESTAURANTS

For several years there has been a trend to top off high-rise commercial and office buildings with a revolving restaurant. This consists of a circular floor section that carries the dining tables and is enclosed with glazed walls. It rotates on rails around a stationary core housing the kitchen and service facilities. The restaurant generally rotates at a rate of one revolution per hour. Statistics have shown that most patrons like to stay for at least one complete revolution to enjoy the view. The speed of rotation is adjustable and is usually increased slightly when there are many customers waiting in line. The higher speed, of course, permits higher usage of the facilities and increases business.

The floor has a structural steel frame supported on wheels which are designed for silent operation and are driven by an electric motor. Because of its access to the outside air, the frame requires good insulation and in cold climates the rails may require de-icing facilities or provisions for heating the underfloor space. The glazed walls should have thermopane or other double glazing. Some designs have considered heated electric defrosting wires built into the glass, a rather expensive solution both in initial cost and energy consumption.

GRAPHICS AND DECOR

The visual impact of a commercial building has been related psychologically in part to its commercial success or failure, a fact well known to experienced shopping center developers or operators and to those in related professions. This visual impact, apart from that of the overall design of the building, includes such factors as window design and arrangement, storefront and door design and finish, and hardware. Among the major visual factors are graphics and decor. These include such items as planting, artwork, special features such as seating areas, fountains, and so forth and all items including written or visual signs that give directions and other information to the public. Wherever possible sign language using symbols should be used. Certain symbols are fairly universally used, for example, the signs for washrooms and wheelchairs. Telephones, escalators, elevators, and so on are easily indicated by corresponding symbols. (See Figures 3-44 through 3-49.) Store signs in a mall must be controlled and specifications must be enforced regarding restrictions. All signs should be compatible with each other and in good taste. They must also be in keeping with the building decor. Supergraphics can often be used to emphasize certain building areas or features, but they should not be too extreme or gaudy. Since tastes and fashions tend to change, it could be expensive to change items that are in vogue for only a short time. This could include individual signs for specific tenants, direction indicators for features such as kiosks which may disappear in the course of time, or other building areas which are meant to be emphasized.

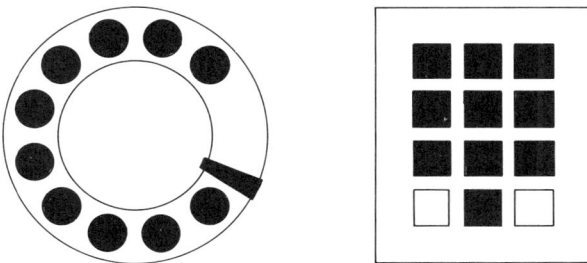

FIGURE 3-44. Telephone.

FIGURE 3–45. (a) Men's washroom. (b) Ladies' washroom.

FIGURE 3–46. Handicapped persons.

FIGURE 3–47. First aid.

FIGURE 3–48. Food facilities.

FIGURE 3–49. Cloakroom.

FIGURE 3–50. Public transportation.

FIGURE 3–51. Taxicabs.

FIGURE 3–52. Escalator.

FIGURE 3–53. High tension.

COLOR CODING

Color coding is often used to identify piping, equipment, or materials that have special properties or special purposes. In buildings there are various systems which should be easily recognized. Although the patterns may vary, there is a somewhat standardized color scheme as follows:

A. Red: Used for items concerned with fire protection and fire fighting, sprinkler installations, hydrants, fire extinguishers, etc.

B. Green: Identifies safe and harmless materials which are non-flammable, are not poisonous, toxic, or explosive and are not under high pressure or at high temperature.

C. Yellow: Indicates materials of a dangerous nature, which may be highly volatile, flammable, toxic, poisonous, or corrosive or under high pressure or temperature, including fuels, explosives, etc.

D. Blue: Stands for materials which are used to counteract poisonous substances (solid, liquid, or gaseous) or which neutralize certain chemicals.

E. Purple: Denotes valuable substances requiring careful handling and conservation.

FOOD FACILITIES

Many buildings other than shopping centers and other commercial buildings housing full-fledged restaurants have snackbars or other rudimentary prepared-food facilities, primarily for the convenience of the building population. These facilities require certain provisions and services which should be included in the initial design, if possible, since adding them later can be a very costly and technically complicated matter.

Water and drainage are usually not too much of a problem. Additional electrical capacity can also be provided relatively easily because most of the equipment is comparatively small and does not draw excessive power. What could create a problem are the ventilation and air conditioning ducts which require space and which at times may be very difficult to find at a later date. Provisions also have to be made for garbage areas, which under special circumstances may even require refrigeration. Depending on the volume of business, a cold storage area as well as regular storage space may be necessary. It is advisable to provide sufficient refuse receptacles on the premises or else the surplus refuse will be found on the floor. Some local codes may insist on public washrooms for men and women, which, in addition to the technical problems they could create, could become an expensive item.

SERVICE AREAS

Service areas in commercial or office buildings are generally the non-leasable areas. In shopping centers they are usually considered as a separate category. Included in service areas are boiler rooms, transformer vaults, valve chambers, meter rooms, miscellaneous mechanical and electrical service rooms, washrooms, corridors, passages, stairs, storage and loading areas, garbage rooms, and facilities for building staff.

Since service areas are considered as financial ballast, there is a tendency to design them to minimum requirements, a policy which often backfires when additional equipment has to be installed at a later date. Leaving some additional space to spare can be very convenient in the future and can prevent many problems at relatively small cost.

In high-rise buildings, service areas are usually concentrated in the building core. Pipe spaces and elevator shafts often form a shear wall that acts as a spine for the building.

Depending on the nature and purpose of the building, the washrooms may be kept locked, as is usually the case in office buildings, or open to the public, as is more likely the case in commercial buildings. This has often caused problems in shopping centers where washrooms are sometimes abused, sometimes by drug addicts, and at other times are vandalized. It is advisable to have lockable hardware on washroom doors.

Long passages and corridors, especially in the basement areas, should be avoided because they tend to be used more for storage than for traffic and can become a fire hazard.

Finishes in service areas should be in keeping with their use. Elevator lobbies and main passages should have first-class finishes to match the quality of the building and to allow for high wear and tear. Service corridors and areas should be designed for heavy abusive traffic. Floors should have a hardened metallic cement finish, particularly in shipping areas, and walls should have protective dadoes unless they are made of heavy masonry or concrete. Exposed pipes and corners should have protective guards, as should exposed sprinkler heads and electric fixtures which could be damaged by forklift trucks or other equipment.

The staff rooms for building maintenance people and cleaners should have sufficient electrical outlets, a common design deficiency in such areas. The following outlets may be required: utility outlets on work benches, outlets for battery chargers for cleaning machines, outlets for public address systems, outlets for telephone installation, and outlets for security and alarm systems.

GARAGES

Most building codes require provision for a certain amount of parking area for a building. The parking area is usually expressed as a percentage of the building's floor space and usually varies between 10% and 20%. This space may be inside or outside the building. If it is within the building, the garage has to consider the various factors that affect the rest of the building, and vice versa. Most garages are multi-story garages. This means that they require ramps (usually an entrance ramp and an exit ramp). The ramps must have sufficient vertical clearance, must not be too steep, and must not be curved too tightly. Column spacing must be adequate to permit cars to move easily for turning and parking. Generally, this means a clear space of about 27 ft (8.25 m) minimum between columns. This may be all right for commercial or office buildings, but it can be an awkward module for apartment buildings. Since the area is enclosed, it requires an efficient ventilation system to evacuate exhaust fumes. This may be covered under local codes. Heating and lighting are usually only nominal since they are of lesser importance. For security reasons, garage doors in an unattended or unsupervised garage should be equipped with key-operated door operators in order to prevent unauthorized persons from opening them.

Parking garages are sometimes used as ancillary buildings or as independent self-contained structures. In the latter case, they may be drive-in garages or automatic parking garages in which the cars are placed either automatically or semi-automatically by attendants. These garages are usually built as open structures and may only be equipped with minimal services. There may be no heating system and there may be ventilation only if required by applicable codes.

Lighting is kept very low and may range between 10 fc (foot candles) and 25 fc.

Since these structures are not restricted as to column spacing, they may have varied spans which may also be a function of the type of construction. If precast or prestressed members are used, the spans are often unequal at right angles and are designed to permit free movement of cars, but with maximum spans in the other direction.

Other designs have circular spiraling inclined floors and ramps near the interior service core. Here the cars are parked at an angle to the exterior perimeter.

If the garage is above a certain height, a passenger elevator is required. The elevator should be designed to handle wheelchairs, if necessary, but stairs must be provided in any case. In addition, ancillary facilities may be required, for example, washrooms for customers and office and service facilities for the administration and operating personnel in paid parking garages.

The roof space may be used for open-air parking. In some parts of the country snow removal may be required. Because of the restricted headroom and limited floor loads, only light snow removal equipment can be brought up through the building and it may be necessary to clear the snow off the roof with the light equipment to grade level from where it is removed by full-size equipment.

WINDOW WASHING EQUIPMENT

Most modern high-rise buildings with curtain walls have fixed windows that must be washed from the outside. Window washing is generally done from a swingstage which is motorized and supported from a hoist mounted on a trolley that runs on a roof-mounted track around the perimeter of the building. Water and other services are provided from the roof down to the platform. For lateral support, the vertical mullions of the curtain wall are often used as guide rails. Some installations have safety devices that can anchor the platform to the mullions in case of emergency.

COMFORT TRADES

The shell of a building provides shelter against the elements but nothing else. It would be cold in winter and hot in summer, the air would be stale, and it would be dark from dusk to dawn. Water would have to be carried in from somewhere else and sanitary facilities would have to be located at some distance outside.

In the course of time humans learned to build fireplaces and hearths, put openings in walls to let air circulate, carry water in through pipes, and use candles and lamps for light. It is only within the last two centuries that special trades developed to improve the indoor environment. These trades, which belong to the "mechanical trades," are also known as the *comfort trades*. Although they are specifically concerned with temperature and air quality in a building, they are also concerned with sanitation and lighting. Some, but not all, mechanical trades have certain aspects which concern human comfort and can, therefore, be included under this classification. They include heating, ventilation, air conditioning, refrigeration (when it is part of air conditioning), plumbing, and — although not a mechanical trade — electrical installations. All of these are described briefly in the following sections.

HEATING

Heating systems come in a large assortment, and with the energy crisis worsening variations are appearing continuously, some representing improvements on existing systems and others being new concepts and developments trying to tap untouched sources of energy. The greatest variety seems to have emerged in residential housing.

The basic and oldest system is the open fireplace which in some countries has survived unchanged until today. It is interesting to note that in houses equipped with a proper mechanical heating system addition of a fireplace has become somewhat of a luxury item.

Among residential heating systems, excluding portable fueled or electric heaters, we find the following:

A. Warm air systems, gravity or forced air

B. Hot water systems

C. Electric heating

Some of the smaller houses, especially $1\frac{1}{2}$-story houses, often have a hot air furnace centrally located in the basement and a grill in the ground floor through which the heat rises and distributes itself in the house, also heating the upper part by air rising up the stair. Sometimes there are also floor grills in some locations. If there are any closed rooms upstairs, they may require portable heaters.

Among the more popular systems for houses is the forced warm air system in which ducts run from a furnace in the basement to the different rooms. At the end of each duct is a baseboard register or floor grill, generally under a window, to provide a warm air curtain. The furnace, oil burner, and fan often come as a packaged unit and are easily installed and connected. A humidifier built into the plenum chamber on top of the furnace ensures that the air is not too dry in winter. Return grills are usually located at some height from the floor to ensure air circulation and prevent dead air pockets. A system like this may be adapted to air conditioning.

It may be of interest to note that a rule of thumb, one that is based on statistics, shows that the oil consumption for a medium-sized house in a temperate zone averages about ½ gallon per square foot of floor area (45 liters per square meter).

Some larger houses have a hot water system. In older houses the hot water circulates through cast-iron radiators, but in modern installations baseboard convector units are used in most cases. Circulation pumps are essential in this kind of installation. It is necessary to ensure that no air pockets are left in the piping and that the high points are provided with vents to enable bleeding any air trapped in the system.

In some housing projects that feature all-electric houses the heating system consists of built-in electric heating units. This may represent a great saving in initial cost for the builder, but it can be extremely costly for the future homeowner, particularly if the local electric authority has a demand rate electric tariff system. In a case like this, the best solution for the homeowner is to install some kind of meter miser. This device ensures that whenever a high electric load appliance is turned on no other appliance can be turned on at the same time. It prevents an electric peak load on the system, which could result in an inordinately high electricity bill.

In some of the more expensive houses and in the southern regions of the United States many houses are equipped with complete air conditioning systems or combination heating/air conditioning systems.

Most of the heating systems are either oil- or gas-fired. Although the general fuel situation indicates a potential return to coal, coal will be used in large buildings rather than in homes.

One of the more recent developments is solar heating. This system includes the following components: sloping solar collectors on the roof oriented toward maximum sun exposure. These collectors are metal boxes painted black and they are covered with glass panes to keep the heat in, similar to the action in a greenhouse. In the boxes are copper coils that are filled with water. The copper coils are heated by the sun. The water is drawn or pumped into a large storage tank, usually in the basement, which acts as a heat storage reservoir.

So far, although the technology is well advanced, the economic aspects have prohibited these systems from being competitive enough, but this may be only a question of time.

A hot air version of solar heating panels is still in an experimental stage. Instead of hot water pipes, this system is based on circulating hot air, which has been heated in the slanting roof boxes, through ducts over a box filled with rocks which heat up and act as the heat storage reservoir. In the hot water version the water is circulated through the heating units of the house by means of a circulation pump, but in the hot air version a fan blows the hot air through the ducts.

It must be noted that in most locations solar heating, even under optimum conditions of sun exposure, may not be able to supply 100% or a steam heating system, either gas- or oil-fired. (In the not too dis-involving a separate secondary heating system, may be required.

Commercial and industrial buildings usually have either a hot water or a steam heating system, either gas- or oil-fired. In the not too distant future they may return to using coal.) Of course, in regions where there are nuclear power stations the heating systems may be all-electric installations but these may prove, although less expensive in initial installation costs, very expensive in operating costs on a long-range basis, since the cost of electricity, especially on the basis of peak demand loads, is skyrocketing.

Buildings that have combined summer/winter heating/air conditioning systems generally have heating coils in the fan coil units. These systems are thermostatically controlled. Whereas air conditioning must be controlled by zones, some heating units, depending on the system used, can be controlled individually and thus can take care of localized cold spots.

Usually there are areas in a building that require special provisions for heating, for example: entrance vestibules, garages, closed passages, stairs, washrooms, loading docks, and shipping areas. These areas are generally serviced by independent unit heaters.

Shopping centers, especially the mall type, have comparatively low heat loads and often they do not require more than auxiliary heating systems. In all-electric centers, electric unit heaters may take care of nominal heating requirements when certain important sources of heat are not available. These include the heat generated by the lighting from stores, show windows, and the mall, and the heat generated by people in the center. This usually occurs on weekends when the center is closed.

Under certain conditions, radiant heating systems may be used. This system consists generally of hot water heating coils located in the ceiling or walls and which radiate heat into the premises. Another radiant heating system consists of pipes embedded under the paving in a driveway or plaza to melt snow in winter. This system can be costly and if it springs a leak, it can be awkward and difficult to repair or replace.

In principle, the design of a heating system includes replacement of the heat lost from the building. This may occur by conduction and radiation through floors, walls, and ceilings and by leakage through exterior openings. Thus the more insulation there is in ceilings, walls, and floors and the better the exterior openings are caulked and weatherstripped, the less heat will be lost and have to be replaced, resulting in a smaller heating system and commensurate savings.

The heat loss, among other factors, is a function of the difference between interior and exterior design temperatures. Since the maximum low temperatures last for a relatively short time only, it would be uneconomical to use them as a basis for design. Therefore, a slightly higher temperature is used. Exposure to high winds may also increase the heat loss of a building.

Special care should be given to the location of the fill and vent pipes of oil tanks. These pipes must be accessible to the oil trucks for filling and because of the possibility of spillage they should be kept away from sidewalks, planting areas, and places of high visual exposure.

VENTILATION

Although ventilation may be combined with air conditioning, there are certain areas in buildings that require separate ventilation systems.

Garages usually require extensive ventilation to get rid of the exhaust fumes.

Restaurants require exhaust systems in their kitchens to get rid of heat and cooking odors. Since such a system can carry grease in the fumes, the ducts must be cleaned periodically to prevent grease fires in the ductwork.

Dry cleaners require ventilation because of the fumes from cleaning solvents; in fact, some local regulations insist that customer areas be closed off from work areas.

Washrooms, corridors, stairs, and other enclosed areas require exhaust systems. Corridors and stairs are often slightly pressurized to prevent smoke and flames from spreading in case of fire. Ductwork requires automatic fire dampers for protection. When the air or fumes to be exhausted are localized, it may be of advantage to install a hood to collect them, for example, over a charcoal broiler in a restaurant.

In certain plants that use process steam, ventilation is required to remove the excess humidity to prevent damage to the building, or mildew.

Rules on air changes vary greatly and depend on prevailing conditions. Make-up air taken from outside may require heating or cooling and humidity control. A ventilation system generally will exhaust air and introduce fresh air from outside the building. This air will be at the temperature of the outside and will contain the amount of vapor presently in the outside air. If this air is treated, either by heating or cooling it, and/or by adding moisture or extracting it, the ventilation system will become an air conditioning system. If air is recirculated inside, it requires some treatment. Offices are often based on about 10 air changes per hour, but this number increases rapidly in restaurants or manufacturing plants. People doing hard physical labor or involved in strenuous activities require a high degree of ventilation, but there is a limit to the number of changes because too many air changes create a high air flow which, particularly to overheated persons, can be very unpleasant.

AIR CONDITIONING

Until a number of years ago air conditioning was considered a luxury. Now there are hardly any buildings or private houses that do not have air conditioning. The popular concept assumes that air conditioning is only required in warmer regions, but the fact is that it may be needed in much cooler regions, either because there is a rather hot, though short, summer season, or because the building has a large heat gain to dissipate which may come from some internal source such as lighting, process equipment or machinery, heat-sensitive installations such as computers, or for other reasons.

Air conditioning has a number of objectives:

A. To clean and purify the air
B. To regulate and control the humidity in the air
C. To heat or cool the air
D. To circulate the air in the building

Generally, the air in the building is recirculated, but in the process of reconditioning it additional fresh air is usually required from the outside.

The air can be cleaned in several ways and for different purposes. Electronic precipitators can remove unwanted dust particles. The quality of the air and the removal of odors can be improved by ionization, by ozone treatment, or by other methods. This is important because a higher relative humidity tends to increase the intensity of odors. If these odors become particularly annoying and objectionable, more fresh air may have to be substituted for recirculated air.

The air handling units take care of the humidity by first extracting excess moisture by condensation. The air and moisture meet in the spray chamber where additional moisture may be added in winter or removed in summer by condensing it on cooling coils. If heating is required, the air can be directed over heating coils or convectors. After it has passed through the various stages, it is circulated throughout the building via the ductwork.

The refrigerants, which may be air, water, or gases such as freon, are compressed first. This compression generates a certain amount of heat which can be eliminated by circulating the refrigerant through cooling towers in which the excess heat is dissipated into the atmosphere. The refrigerant is allowed to expand, at which point it will tend to vaporize, resulting in a cooling effect for the air which circulates past the cooling coils.

Since a large water-cooled installation requires a large amount of water, some regulations insist that these installations recirculate the cooling water in order to reduce water wastage from the municipal water supply. Small units are usually air-cooled and do not have this problem.

There are two broad categories of air conditioning installations:

A. Individual units
B. Central systems

Individual units are used mostly for residential purposes or for individual offices in older buildings that have no central air conditioning. These units are usually window units, but they may be built into exterior walls. They start from a capacity of about 1500 W or $\frac{1}{8}$-ton cooling load and at 110 V can be plugged into the regular house current. Above a certain size, depending on local current characteristics, they may require their own circuits, usually at a higher voltage.

Forced warm air heating systems in some houses can be adapted to air conditioning by adding condensing units, cooling coils, and other accessories, but they do not always result in efficient systems because the ducts may be too small and because the heating registers are at floor level. For air conditioning, the registers should really be higher up if they are to be effective.

Office and commercial buildings have central systems which often are combined with the heating system to give a summer/winter all-year-round system. An office building may have several systems or a combination of different systems. A perimeter system may have fan-coil units around the exterior walls. This system is especially convenient when used with a curtain wall or continuous ribbon windows and it may be installed with a continuous enclosure to sill height. Core systems serve the service core, which is usually at the center of the building. They may also feed the floor by means of ducts leading to ceiling diffusers or by having a ventilated ceiling where the ceiling space becomes a plenum for the system. In this case, the air is diffused either through slots in the ceiling tiles, through suspension, or through combined light fixtures/diffusers. Fire departments and underwriters are often opposed to the ceiling plenum system because if there is a fire, the flames and smoke can easily spread through the ceiling space. When the ceiling is used as a plenum, there are usually regulations that restrict the maximum size. The upper limit is often given as 5000 ft, which necessitates installing smoke baffles in the ceiling space.

Some buildings have to allow for special conditions. In shopping centers the mall and stores usually have different requirements for both heating and cooling. For flexibility as well as for economy, individual roof top units are often used for the stores and a central system is used for the mall and other public areas. This is often dependent on the leasing policy and there are several alternatives. The owner of the center may either install a complete central system or supply cold or chilled water to the store location with distribution the responsibility of the tenant. In either case, the tenant is either charged the proportionate cost in the rent for the store or else is made fully responsible for supplying his or her own air conditioning system. Thus the owner is saved a good amount in capital investment but obviously loses a certain amount of rent revenue. In the latter case, an air conditioning installation in the store is often obligatory on the part of the tenant.

A central installation is generally more expensive than the equivalent number of roof top units. If the installation is above a certain capacity, it may require special personnel similar to the personnel of a boiler installation. In addition, a breakdown would affect the whole building. A central installation is a necessity when individual units cannot be used. A shopping arcade in a high-rise building or a multi-story shopping center would have to have a central installation.

The design of air conditioning depends on a number of factors, including the differential between interior and exterior design temperatures, actual ambient temperature pattern, relative humidity conditions, heat generated in the mechanical systems, and heat gain of the building and contents. Heat gain through windows is part of this consideration, but this heat gain can be partially offset by using solar or reflecting glass or curtains or blinds. Heat gain through walls is of very little significance since the thermal resistance of the walls usually prevents the heat from penetrating fast enough to become a factor during daytime. In a sense, heat gain is analogous to heat loss requiring heating, but it is not necessarily identical. Heat gain can be offset or reduced by flooding or sprinkling roofs to prevent heat build-up in the tar or asphalt. The biggest interior heat gains in office or commercial buildings result from lighting, the people load, and office equipment. People load heat gain varies considerably. For example, people engaged in strenuous physical activities can generate three times as much heat as sedentary people in an office. Although there can be a great variation in the actual requirement, there is a rule of thumb that a ton of cooling capacity under average conditions in a temperate region takes care of about 350 sq ft ($32\frac{1}{2}$ m^2) of office space.

REFRIGERATION

Refrigeration in buildings is used mainly in supplying the cooling part of air conditioning systems or in coolers and freezers for commercial or industrial uses.

Ice is, of course, a natural refrigerant and humans have been familiar with its properties since prehistoric times. Its ability to preserve food from being spoiled by excessive heat has been traced back to a very early era.

Modern refrigeration systems use different mechanical methods to create the cooling effect. For solids to melt to liquids or for liquids to vaporize, heat, which can be absorbed from exterior sources, is required. The refrigerant, which may be a gas, is condensed into a volatile liquid in the condenser which must be cooled because of the heat build-up. From the condenser the refrigerant is routed to the evaporator where it is allowed to expand. In this process the refrigerant absorbs heat from the air to be chilled, which then is circulated through the cooling coils.

Cooling capacity, as usually defined, is expressed in *tons of refrigeration* which is equivalent to 12,000 btu/hr or 288,000 btu/24 hr. This is also the quantity of heat required to melt or freeze one ton of ice (144 btu/lb $\times$ 2000 lb = 288,000 btu). The condensing unit can be cooled either by air or by water. When water is used, local codes may require that the water be recirculated in order to conserve local resources. In installations for private houses the condenser may be an air-cooled unit mounted outside the house.

PLUMBING

Plumbing systems in buildings are basically concerned with water supply and drainage. They may also include distribution systems for other materials. The scope can be divided roughly as follows:

A. Water Supply
  (1) Domestic (potable) water
  (2) Fire protection
  (3) Water for heating and process purposes
B. Drainage
  (1) Sanitary drainage
  (2) Storm drainage
  (3) Disposal of hazardous waste
C. Distribution Piping
  (1) Gas
  (2) Other substances

Domestic Water

Domestic water is usually supplied from municipal services, which means that the water undergoes some treatment which could be chlorination, ozonation, or some other purification method in order to render it potable. The water may also be fluoridated to combat tooth decay, but fluoridation has proved to be a controversial subject in many localities.

Isolated buildings may require individual provisions for their water supply. This supply could come from artesian wells. It is mostly houses such as summer houses in the country which have to rely on wells unless they happen to be located on a lake or river. Since these summer houses also often have a septic tank, it is very important to choose the right location for both well and septic tank in order to prevent contamination. In any case, the water must be treated. Chlorination is one of the most common methods of treating water.

In some places where there is no fresh water from natural sources, water may be collected from rain and other precipitation and stored for later use. Many areas that have only a salt water supply may require desalination plants in which fresh water is produced by distillation. The problem with this process is the inordinately high amount of energy required compared to the yield and the high cost of the water recuperated.

One of the more recent, and rather exotic, schemes being investigated is to tow icebergs from polar regions to some of the arid desert regions such as the Arabian peninsula where water is badly needed for domestic use and for irrigation purposes. Feasibility studies claim that the loss by evaporation or melting during the long journey would be minimal. What may, however, be more of a hazard could be the ocean currents heading the wrong way and pulling both the iceberg and the towing tug with them. Perhaps this scheme could be called pouring water on troubled oil!

Fire Protection

Water for fire protection may also come from municipal sources. It may either be taken from the building water entry or have its own separate entry from the street. The latter is usually incorporated under the (fire protection) sprinkler contract rather than under the plumbing contract.

The sprinkler system may require extra booster pumps to generate the pressure required, especially in high-rise buildings, whereas domestic water may be either boosted with the aid of pumps or by using an impounding reservoir water tank at the top of the building. The water quality in the case of sprinklers is not relevant.

Water for Heating and Process Purposes

If the building has a steam boiler either for heating purposes or for process steam, it usually takes the water from the regular entry, but special treatment may be needed to soften the water to prevent the formation of scale and to prevent the system from corroding.

Corrosion is caused primarily by free oxygen in the water which oxidizes iron and causes rust. Metals have different electric potentials. When dissimilar metals are in contact and become exposed to moisture, an electric current will be generated between them, which sets up what is known as galvanic action. This starts the oxidization process which results in the rust. In the latter case, cathodic protection systems are used to counteract the galvanic action. Anodes are attached to the piping and metal components of the system and energize a countercurrent, all of which combine together to neutralize the corrosive effect on the mechanical system. Oxygen can also be removed by de-aerating the water and recirculating it in the mechanical system with a minimum of fresh water. Too much fresh water would tend to replenish the oxygen content.

Sanitary and Storm Drainage

Most buildings and just about all residential housing have a single drainage system that handles the sanitary drainage from washrooms, kitchens, and so on and the storm drainage from flat roofs. There may be exceptions, for example, major buildings in a municipality that has a dual sewer system and insists that roof drainage be connected to the storm sewer system separately. This could occur in high-density areas where there may be a concern about overloading one of the sewer systems. A double system, needless to say, can add considerably to the cost of the building. Sewer systems are often designed and built long before the area is fully built up. Later these systems may be inadequate to handle the ultimate peak loads imposed on the system. If the municipality is unwilling or unable to enlarge or replace the sewer system, it could install a special reservoir to handle these peak loads until the existing system can run off the excess. It is also very important for the building to have a good back water valve installation to prevent the sewer system from backing up. This is especially important for private houses that have basements. In most localities there is a regulation that requires check valves in the house sewer.

Disposal of Hazardous Waste

Certain industrial plants handle materials which are either hazardous, poisonous, or prohibited by law to be discharged into the regular sewer system. These plants may require special installation either for purification or for alternate disposal. The installation may include a separate piping system that has a collector pit from which the materials must be removed and disposed of independently from the municipal sewer system. Service stations are not allowed to flush either gasoline or oil into the regular sewers and they have special facilities for waste oil collection.

Distribution of Gas and Other Substances

Many buildings in which gas is used for heating, cooking, and air conditioning require a special distribution system. Similar systems from a central facility are sometimes used for specialty installations;

for example, hospitals and medical buildings have systems for bringing oxygen and nitrogen to various rooms and offices. It must be kept in mind that these installations present special hazards and require provisions for shutting off and isolating the system, or part of it, in case of emergency.

FIRE PROTECTION

Fire is perhaps the greatest hazard in buildings both during construction and after completion. There are many different causes of fires and although negligence may be first on the list, especially when it comes to careless smoking, there are many other equally inexcusable causes, not the least of which is sloppy housekeeping on the construction site. Faulty or improperly maintained temporary heating equipment probably comes a close second. Other causes include various careless equipment installations and other mechanical misdeeds.

Three elements are required to cause combustion: fuel, oxygen, and heat. To fight fire effectively, it is necessary to eliminate one of these elements, which is the purpose of fire protection systems. Different materials, however, burn differently and as a result require different fire fighting techniques.

Fires are categorized as follows: *Class A* fires involve materials such as paper products, wood, fabrics, garbage, and other ordinary combustible materials that can be extinguished by using water or certain chemicals. Portable fire extinguishers with pressurized water, loaded stream, or soda acid are suitable for small localized fires. *Class B* fires involve higher hazard materials such as oils, paints, bituminous products, and other flammable liquids. These fires require treatment with foam, dry chemicals, or carbon dioxide. Portable extinguishers can be used to put out small fires. *Class C* fires are electrical fires involving live equipment installations. Only certain chemical or carbon dioxide systems or portable extinguishers are suitable for extinguishing them. *Class D* fires, which generally occur in industrial plants, involve certain metals which generate extremely high heat when burning. They must be handled with dry chemicals because of the explosion hazards.

Until comparatively recently fire protection was more or less optional in building construction. The increase in actual fires, as well as the proportionate increase in costs of fire damage, has resulted in most building codes requiring fire protection systems for most new buildings. Some possible exceptions are housing and certain low-hazard commercial and industrial buildings, but even high-rise apartment buildings may be required to install sprinklers.

Basically, fire protection systems, excluding portable extinguishers, can be classified as sprinkler, standpipe, or chemical systems, with certain subdivisions under each category. Sprinkler systems are automatically activated and standpipe systems are handled manually. Chemical systems may be either automatic or manual, depending on the type and location. A building equipped with sprinklers may also have standpipes which supply water to firehose cabinets. In high-rise buildings in which a fire on the upper floors may be beyond the reach of the fire brigade both physically and in terms of time it is advisable to arrange for building occupants to be trained in fire fighting and to designate special fire marshals to deal with emergencies. This includes establishing fire drills, building evacuation procedures, and other precautions.

Sprinkler heads are usually spaced according to the rating of the occupancy of the premises, but spacing also depends on the exposure. Additional heads must be installed in hidden or shielded spaces. Sprinkler heads vary in types and they can be activated at different temperatures. The system may be either wet (water filling the pipes at all times) or dry (water filling the pipes only when the system is activated). Dry heads must be used specifically in locations which may be exposed to freezing temperatures, such as shipping areas. Generally, sprinkler heads are activated only individually and cover a limited area. For certain hazards, it may be desirable to have a number of sprinklers discharge simultaneously. This is a deluge system in which special preset controls activate all the sprinklers in the designated area.

Depending on the system, the water may have to be shut off manually after the fire is extinguished or the heads may close automatically and be reactivated if the fire flares up again.

Chemical systems are used when water is not suitable. They create different types of foams which must be suited to the location and nature of the hazard. There are other special systems using carbon dioxide, halon gas, or powder-type chemicals.

For isolated small fires, portable extinguishers may be used. It is also a good idea, if building regulations do not require it, to have a number of portable extinguishers at strategic locations on the site during construction.

It should be noted that carbon dioxide is a serious hazard if used in a confined space. In such an area it is safer to use the halon system.

Sprinkler systems are usually equipped with supervisory alarms which may be tied in to the building security system. The alarms indicate the location of the activated sprinklers and thus locate the fire. The alarms may also be connected directly to the local fire department.

In locations where water damage would be a serious matter, for example, in libraries, archives, or stationery vaults, heads which shut themselves off automatically should be used in order to minimize potential damage.

Many building codes require that special fire protection features be incorporated in the building. For example, automatic vents and smoke hatches may be required in stair and elevator shafts, public areas such as shopping malls, and other areas where there could be potential concentrations of people.

It may be necessary or expedient at times to install sprinkler systems in existing buildings. Depending on the nature of the construction of the building, it may be possible to install the distribution piping in the existing ceiling space by inserting prefabricated piping through a number of openings to keep damage to existing ceilings to a minimum. The mains, if possible, could be located in secondary or service areas since they would probably be too large to be hidden in the ceiling. Although this may be very expensive, the resulting savings in insurance rates over the years may compensate for the cost of the installation.

ELECTRICAL INSTALLATIONS

The purpose of an electrical installation in a building is to provide lighting and power.

The electrical requirements of a major building can be substantial. Since the capacity needed may be extremely high, one of the first considerations is whether to have a high-tension entry and a separate transformer vault for the building or to have the entry already on a stepped-down lower voltage. This is strictly a matter of economics since the local power company may charge a very high rate for current which has been stepped down by them. This aspect may also be very important when considering an all-electric building, for it may represent a saving in capital investment by saving on various major mechanical building components, but the saving may be more than offset on a long-term basis by higher electrical rates and the lack of alternatives in case of power problems.

Specifications for any electrical installations are set to very rigid standards. The hazards involved in electrical work warrant special precautions and care, and all installations are under the jurisdiction of special authorities and are subject to special inspections and monitoring.

The lighting load depends on how the premises are used and is perhaps easiest to design for most buildings since a uniform illumination level can be used for the design criteria when the nature of the occupancy is known.

The power load may be another matter. It depends on the requirements for the mechanical installation (heating, ventilation, air conditioning, etc.) and it varies according to the building population density, usage, orientation, type of construction, type of windows, type of lighting, other powered appliances including office equipment, and other factors.

The combined lighting and power load represents the total electrical load for which capacity has to be provided. For smaller buildings or houses, this is a simple matter because the entry can be chosen from a number of standard entries and the only other variable is the internal distribution.

The electrical entry for a building must have a breaker or switch so that the power can be disconnected if there is an emergency. The breaker or switch should be easily accessible to all authorized personnel. The entry terminates in a panel from which the cable may be split up into different panels for further distribution, combined with meters for measuring the different types of consumption.

In many office buildings the distribution of electricity for office equipment and the telephone distribution may be through special floor ducts which afford much greater flexibility for office layouts. Cellular floors are alternate means of underfloor electrical distribution and they save the installation of separate ducts. If there is no underfloor distribution system, the alternative is to have the wiring in the ceiling and have outlets on special poles installed at strategic locations. This system, however, often has a tendency to result in octopus wiring which is neither attractive nor advisable.

In the case of major buildings and certain specialty buildings such as supermarkets, emergency systems are often installed in the form of emergency generators which cut in automatically as soon as the regular power supply is shut off or cuts out. The capacity of this system is usually restricted to supplying power to stair shafts, selected elevators, and strategic locations on floors to permit evacuation of the building.

This section deals primarily with the mechanical movement of people and goods in buildings and in this context includes elevators and escalators, freight elevators and dumbwaiters, moving ramps, man-hoists, and conveyors. Automobile elevators in parking garages are a specialty item in specialty buildings, as are automotive hoists in service stations.

There are a number of characteristics by which different firms distinguish elevator installations. These characteristics include cabin size, speed of travel, cabin finish, and type and arrangement· of motive power. But the main distinction is in the type and nature of the controls. In large high-rise buildings, specifically office buildings in which there are large in-house and transient populations and multiple elevator banks, controls can become extremely complex and require highly sophisticated computerized equipment to schedule elevator traffic efficiently. This traffic must allow for rush-hour peak loads up and down, which means that there must be separate programs for morning and evening traffic. Intermediate and intermittent traffic during business hours requires a different program to ensure that waiting time is kept to a minimum. Elevators are usually divided into traffic zones and they may serve specific floors only. The number of floors ranges from 10 to 20 and depends on the overall number of elevators installed. Some elevators may be used as express elevators that stop only at specific floors. Other floors are reached by local traffic elevators that service a limited number of floors. At times this system can create a shift in waiting time from the lobby to an upper floor.

In some buildings the elevators have two-story cabs. One level of two-story cabs usually serves even floors and the other serves odd floors, a fine arrangement, except in buildings in which because of superstition, the thirteenth floor is omitted. Instructions on how to use the two-story cab should be posted on every floor.

A somewhat recent trend is to mount glassed-in elevator cabs on the outside of buildings or in very high atrium spaces. It may be advisable to have conventional elevator cabs too, since some people, especially the elderly, tend to get dizzy when exposed to heights.

Elevators are either hydraulic or electric. Hydraulic elevators are restricted to a certain number of floors and speed of travel. Freight elevators in plants or low buildings often are hydraulically moved. The motors for electric elevators are usually in a penthouse above the shaft. The cab runs on special guide rails and is equipped with an emergency brake device which can clamp onto the guide rails. The cab is connected by wire ropes to counterweights which run on auxiliary rails at the back of the shaft. The ropes pass over sheaves in the penthouse connected to the motor. At the bottom of the shaft there are safety spring-loaded bumpers to take the impact of the cab if it should not stop properly at the bottom floor. The system used for roping the elevator may be of the single or double wrap type. In the former, the ropes pass over the drive sheave once and are wedged into the sheave grooves, thus supplying the necessary friction. In the latter, auxiliary idler sheaves are used. In this case, more rope is required and the mechanism has to pull more weight in proportion.

Elevator doors of small cabs should be wide enough for two persons to enter at the same time, but in office buildings they may be twice that size or more. Elevators in hospitals must have enough space in them to accommodate moving beds from one floor to another. Freight elevators often have vertical bi-parting doors to save on the width of the shaft. It is absolutely essential to ensure that the shaft is perfectly straight, especially in a high-rise building.

As mentioned before, elevator controls in office buildings can be very complex, but few apartment buildings have anything other than simple controls. As a result, there is a long waiting time for elevators. The latter is strictly a matter of cost since better controls would expedite the movement of people and reduce the waiting time at different floors. Better controls would also tend to reduce the peak loads which are a result of a cheaper elevator control system. In freight elevators constant push-button controls are often used. In these the cab moves only when the button is held down and it stops whenever the button is released. In buildings in which there are several elevators at least one elevator should be prepared for the use of the handicapped and for people in wheelchairs. The control buttons should be within their reach.

Basically, escalators are automatic stairs mounted on inclined trusses between different floor levels. They are reversible and are used in opposite directions in some locations during morning and evening rush hours. The only adjustment they require is that for speed, but there is a maximum safe speed which is limited by applicable codes and for safety reasons cannot be exceeded. Under emergency conditions and in case of power failure, escalators become the equivalent of ordinary stairs and allowance for this may be made in the requirements for emergency exits for the building. Location of escalators is important because elevators tend to create traffic buildup and local congestion. This can be offset by using features such as kiosks or planters in malls to direct traffic in the desired direction and density. Because escalators have an inherent hazard from clothing or other items which might get caught in the treads or guard, the escalators should have an attendant near them during peak use periods. Unfortunately, few places bother taking that precaution.

Moving ramps are longer and have gentler slopes than escalators, consisting of long sections which may be horizontal, or have a shallow slope, or consist of a combination of horizontal and sloping sections. They are sometimes used at airports to save people walking long distances to departure lounges. Because they have a flexible surface that moves over idler rollers, ramps have a slightly uneven surface which causes some people to lose their balance. Ramps may be used in department stores to enable people to move shopping carts easily between floors, but they require a lot more floor space.

Dumbwaiters, which are small elevators, may be motorized or pulled by hand on a rope. They are used for moving food in kitchens or restaurants or for moving papers in offices or banks. Because they are in shafts in buildings, dumbwaiters may require fire-rated shafts and doors.

Conveyors are used on a limited basis in buildings in which there is no freight elevator. They are primarily used to move goods into a basement or, as in the case of supermarkets, from the store to a car order pick-up. Conveyors are usually motorized and are reversible.

Man-hoists, or paternosters, have specialized use in multi-story garages in which attendants park the cars. The man-hoist is used to bring the attendants back faster to the starting point. Because they can be dangerous, their use has been prohibited in many places.

4

Contracting: Planning, Methods, and Controls

PROJECT MANAGEMENT AND CONSTRUCTION MANAGEMENT

For quite some time now the terms *project management* (PM) and *construction management* (CM) have been used to denote a project execution that is different from the conventional design and contracting methods. In fact, the two terms have sometimes been interchanged, but there is no universally accepted definition, and different people have decided on their own interpretation of each. Perhaps we can analyze the origin and purpose of both PM and CM and clarify some of the existing misconceptions.

Going back a number of centuries we find that when a new building went up the creative force behind it was usually centered in one person who combined the talents (in modern terms) of designer, architect, consulting engineer in various disciplines, builder and contracter, and decorator. In other words, the person created a total project from concept to implementation. Since financing was usually taken care of by a sponsor, usually the local current ruler, the artist often was able to work under conditions in which time and money were of little importance. The stress was on the building's appearance and visual esthetics, the building's utility, and the building's becoming a lasting monument to the builder and/or the sponsor.

In the course of time, many changes took place. Economy became a prime factor. As buildings became more complex, so did the methods to conceive and build them. The concept of specialization began to affect all phases of the project. Design was performed by specialists. Later these specialists became known as architects and consulting engineers. Still later consultants specializing in specific areas came into existence. Construction was implemented by special-ized builders who eventually became the modern-day general contrac-tors and subcontractors. Financing was generally the responsibility of the owner of the building. Thus the standard procedure that gradually developed was to design a building in full, call tenders, and then con-struct the building. All of this was generally based on a lump-sum contract. When certain items could not be finalized before tender call, either a cost-plus arrangement was made or a PC (provisional con-tract) allowance was included.

Certain marked disadvantages of the conventional methods were probably responsible for the emergence of PM and CM. for example, changes on a lump-sum contract became very expensive as well as very time-consuming to process and implement. The contractor had to wait for the completion of the plans before being able to tender on the project. This, plus the emergence of complex projects such as shopping centers which could not be completely designed ahead of time, sparked the ideas to combine and coordinate the work of the designers and builders in such a manner that a project could be built as it was being designed. The theoretical savings in both cost and time would be fully controlled and an acceptable standard of quality would be ensured. PM would take care of the overall aspects of the project and CM would generally confine itself to construction. Below are our definitions of PM and CM:

- The objective of project management is the coordination and implementation of planning and controlling all related services for the design and construction of a project under the optimum combination of the three parameters of best quality, lowest cost, and shortest time.
- The objective of construction management is the coordination and implementation of planning and controlling all related services for the construction of a project under the optimum combination of the three parameters of best quality, lowest cost, and shortest time.

The first aspect of any project is integrated planning from which the project emerges. This planning must be a coordinated effort in which the project manager takes the lead. The second aspect involves setting up the systems and controls to keep the project on track while the design and construction proceed. It must be noted that construc-tion management may be limited in reaching its objective because of constraints imposed by outside factors. For example, if construction management does not have a say in design.

There are certain aspects which are common to both PM and CM. Quality depends on the initial choice of materials and equipment and the workmanship in the field. Cost and time require scheduling and estimates as well as a continuous monitoring system which not only indicates the status of the project to date but also permits taking remedial action if costs or project duration get out of hand.

If the project cannot be given out to one contractor, it is generally broken up into a number of self-contained packages which are tendered on and contracted out on an individual basis. This permits closer follow-up and, as a result, better control. These packages are integrated under a master plan (Figure 4–1).

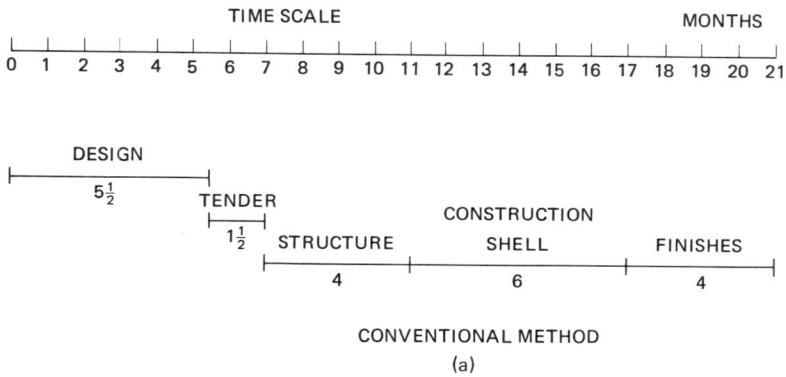

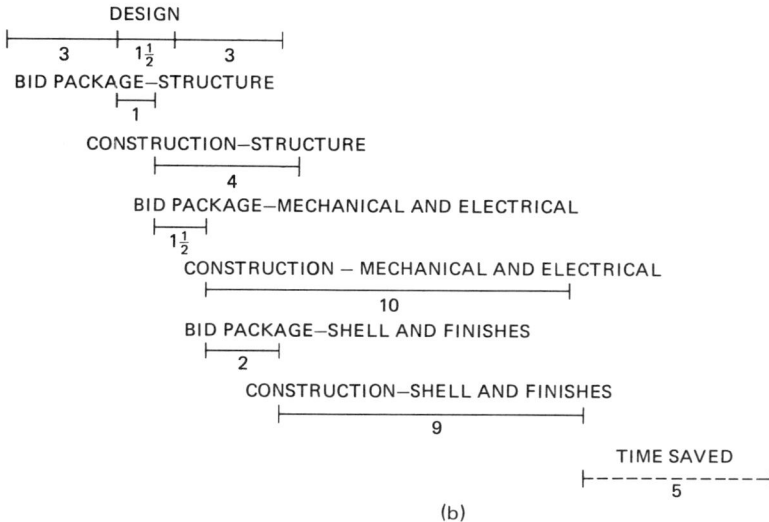

FIGURE 4–1. (a) Conventional method. (b) design - build method.

Many related activities are treated similarly by most firms, but in PM and CM the methods of procurement (procurement of equipment and materials and possibly the procurement of contracts and subcontracts) may vary considerably. Some of the larger firms tend to keep the procurement function in a separate department but smaller firms, and often general contractors acting as project or construction managers, do not separate procurement functions from their usual contracting or purchasing activities. There are arguments for and against both concepts. Some very large and complex projects, particularly those at remote sites, may warrant the additional organization. Arguments against emphasize the cost of the additional organization and the possible duplication of services.

The services provided by PM and CM depend on the requirements of the owner. Sometimes an owner has his or her own PM or CM organization. If this is the case, the organization of the project must be adjusted to suit existing conditions and it should be designed so that there will be a mutual interdependent relationship between the owner, the consultants, and the contractors.

One of the important questions of PM and CM deals with responsibility. Thus contracts given out by PM or CM would be their responsibility. Most PM or CM firms prefer to do their own preliminary negotiations and then recommend potential contractors to the owner. The owner actually awards the contract and thus assumes ultimate responsibility for the contract. Only the responsibility of administration is left to PM or CM. These points must be clarified very precisely in the PM or CM contracts with the owner in order to prevent serious arguments at a later date.

Projects handled under PM are also designated as design/build or design/procure/build projects to distinguish them from conventional contracting methods. The projects may be managed in a number of ways. The project could be on a turnkey basis in which the PM team is relatively independent of the owner. If a target price format is used, the PM team is granted much leeway and latitude. If the information available is better defined and more detailed, more conventional methods can be applied, for example, lump-sum or cost-plus arrangements with or without unit price options. All these are described in more detail in other sections of this chapter.

It should be noted that even though a project is carried out by PM or CM, there is no automatic guarantee that the project will be of the best quality, that it will save money, or that it will take less time to implement. It all depends to a great extent on the skill and integrity of the firm that provides these services. A reputable and experienced firm ensures that the owner's interests are protected at all times and that the owner receives the best value for the fees paid for the services.

The checklist below indicates typical services and the responsibility for these services for PM and CM projects.

Checklist

PROJECT MANAGEMENT
AND CONSTRUCTION MANAGEMENT SERVICES

Note: The services listed below range from partial assistance to full control and from an advising capacity to total responsibility.

Services	*Owner*	*PM*	*CM*
Setting up list of services and degree of responsibility of PM and CM	X		
Financial or mortgage arrangements	X	X	
Economic and technical feasibility studies		X	
Setting up project criteria	X	X	
Commissioning consultants	X	X	
Planning project		X	X
Scheduling project		X	X
Schedule controls		X	X
Budget and control estimates		X	X
Tender documents and tender calls		X	X
Negotiating and awarding of contracts and subcontracts	X	X	X
Purchasing and procuring equipment and materials	X	X	X
Setting up drawing controls		X	
Coordinating design and consultants		X	
Monitoring capital cost management		X	
Maintaining project records		X	X
Tenant coordination for commercial projects	X	X	
Accounting and job control systems		X	
Monitoring construction activities		X	X
Supervising and inspecting construction		X	X
Control of quality of workmanship		X	X
Control of manpower		X	X
Correction of deficiencies		X	X
Processing progress claims	X	X	X
Processing changes		X	X
Finalizing project accounts		X	X
Commissioning of project		X	X

CONTRACTS WITH CONSULTANTS

The first step in creating a project is to organize a design team. To this end the owner must enter into contracts with his or her consultants. It is very important to lay out the consultants' duties and responsibilities in great detail and to settle their fee structure right at the outset because the project may suffer if the consultants do not give the project adequate care and control because they are not sure what their duties and responsibilities are and they are not sure how much they will be paid.

Professional fees are often regulated by professional organizations. The scale of fees (either by law or by negotiation) may vary for different portions of the work. Thus there may be a higher fee for the design of a building than for the design of the site services.

Since the duties and obligations may vary considerably, they should be spelled out clearly. The following should be included in the scope of services and items to be defined:

A. Preliminary work:
   drawings, economic and technical studies, preliminary estimates, reports and recommendations

B. Working documents:
   plans, specifications, tender documents, budget and control estimates

C. Outside consultants:
   engaging and coordinating other consultants, soil tests, laboratories, inspection services

D. Tenders and contracts:
   calling for and receiving tenders, collaborating with owner in awarding contracts

E. Project supervision:
   monitoring and checking work progress, quality of materials and workmanship, adherence to plans and specifications, periodic inspection and progress reports, arranging for testing and reports

F. Job meetings:
   chairing job meetings, minutes of meetings, coordinating other consultants, general contractor and subtrades

G. Authorities:
   liaison with authorities, obtaining building permit and easements, approvals

H. Commercial tenants:
preparing leasing and rental plans, tenant grid layouts, coordination of tenant plans and work

I . Shop drawings:
prompt coordination and approval

J . Progress claims:
processing progress claims, approving progress payment certificates

K. Change orders:
coordinating and processing changes, extras, and credits, adjusting budgets

L. Administrative matters:
extra drafting and design time, traveling and out-of-pocket expenses, additional sets of drawings, "as-built" drawing records, microfilming, telecommunications, reproduction, professional insurance, contract cancellation procedures

Checklist

QUESTIONNAIRE

Company: _____   Location: _____
Nature of Business: _____
Fire hazard: low: _____   medium: _____   high: _____
Waste products requiring disposal:
solid: _____   liquid:_____   gaseous: _____
Process steam required: _____   quantity: _____
Process water required: _____   quantity: _____
Process power required: _____   quantity: _____
Temperature control required: _____
Moisture control required: _____
Corrosion control required: _____
Building areas: plant: _____   floors: _____
                    offices: _____   floors: _____
                    total: _____
Clear height required: _____
Railway siding required: _____
Truck docks required: _____
Elevators required: _____   capacity: _____
Plant floor loads: _____   finish: _____
Personnel: plant: men: _____   women: _____
               office: men: _____   women: _____
Executive offices required: _____
Special provisions required for: _____

Remarks:

PLANS AND SPECIFICATIONS

All regular building operations are based on plans and specifications, which are first used to prepare tenders in contracting work and which later become integral contract documents on award of the contract. These documents, depending on their specific nature, may be prepared by architects, engineers, or special consultants and, depending on applicable codes, may have to be signed and authorized by members of relevant professional associations if they are to be legally valid or suitable for permit purposes. Subtrades use the information contained in these documents to prepare shop drawings for the various building elements.

In order to prepare the relevant criteria for the design, it may be advisable to prepare a checklist of important data. This could take the form of a questionnaire, a sample of which can be found in this chapter.

There are different kinds and grades of plans and specifications. They may be only outline plans and specifications which are generally only used for preliminary and budget purposes and give only rudimentary information. They may be performance specifications which indicate the requirements demanded of various building components but which leave the actual choice to the contractor, or they may be of a generally descriptive nature which are more likely to be used for tender and contract purposes. In this case, the specification writer either provides many details of a general nature or uses material prepared by manufacturers for specific items. The latter, because of its restrictive nature, is only advisable when a specific item or manufacturer is to be used. In such a case, any failure may cause arguments about the responsibility, since the choice was enforced.

The quality of the tender documents has an important bearing on the manner in which contractors prepare their bids and on how the project progresses right from the start. Poor plans and specifications usually result in problems on the project. This is partly due to the fact that many contractors, in view of a general competitive situation, try to take advantage of errors and omissions and often attempt to convert them into extras. This practice results in delays and ill will.

Specification writers often face the question of equivalents. A specified item or brand name may sometimes be substituted by "or equal" or by "or approved equal," usually at the discretion of the consultants. If a cost or delivery time difference is involved, it can lead to serious differences of opinion and arguments. The best way to avoid this is to give specific alternatives.

"Weasel" clauses should be kept out of specifications because they

152

are the mark of unprofessionalism. They include such phrases as "items not shown on plans but required shall be included at no extra cost."

Whenever an error is discovered in a tender document, it should be brought to the attention of the consultants before bids are submitted. Aside from the ethical aspects, this can save a lot of aggravation later. Ambiguities in the documents should be clarified before they become bones of contention.

In order to save additional work, consultants sometimes try to write specifications in as general a manner as possible so that they can adapt them from one project to another. The pitfall here is the possibility that certain items or conditions which are not applicable or suitable for a specific project may become part of the project anyway. In such a case, it is usually the contractor who ends up at a disadvantage.

There are many formats that can be used to prepare specifications. Some organizations such as the (American) Construction Specifications Institute or the (Canadian) Specification Writers Association have published proposed formats for building specifications. Both the American and Canadian Construction Associations and various professional architects and engineers organizations have come up with different formats for construction contracts and subcontracts.

A few general rules for preparing plans and specifications are given below:

- Plans should be clearly and neatly drawn.
- Scales should be adequate and if the building is too large, it should be broken down into sections that clearly relate to a master plan.
- Details should be drawn to a large enough scale to clarify them.
- Dimension lines should be clear and should leave no doubt as to what they refer to.
- Specifications should complement the plans, and vice versa, and items on plans should be explained in sufficient detail. Specifications should cover the full scope of the work required for the project, the workmanship, and the methods and standards expected from the contractor.
- References to standards should be based on those of official organizations or authorities, for example, Urban Land Institute (U.L.I.), American Institute of Steel Construction (A.I.S.C.), Engineering Institute of Canada (E.I.C.), etc.
- Specifications should be written concisely and precisely.

- Specifications should be to the point and avoid unnecessary verbiage.
- The descriptions of items should be technically correct and unambiguous.
- There should be no "weasel" clauses.
- Related items should be combined in appropriate chapters; items should not be buried in unrelated chapters where they could be easily missed.
- Chapters should be as complete as possible (for example, roofing guarantees should be under roofing).
- Headings should be clear and not misleading.
- Whenever there are listings of specific items, the lists should be complete.
- Listing too many cross-references should be avoided.
- When naming specific brand names, several names should be given so that there is a choice. Phrases such as "or equal," "or equivalent," "or approved equal" should not be used.
- General conditions should be presented so that they conform to similar clauses in the contract form to be used on the project.
- Specifications should be printed on fairly heavy paper, should have strong, but not necessarily stiff, covers, and should be bound by means of multiple fasteners. Staples should not be used because the whole specification booklet would have to be taken apart in order to remove individual pages.

Plans are essential in the construction business. They are used to evaluate a building during the tender stage and they are used at the construction site while the building is being erected. They depict the three-dimensional aspects by means of two-dimensional projections.

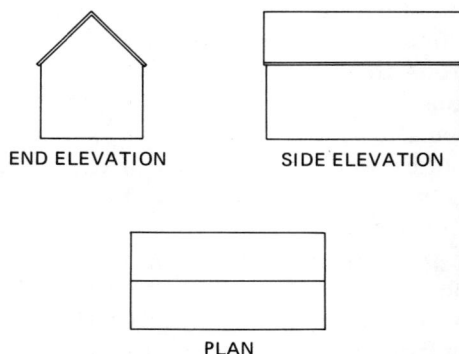

END ELEVATION   SIDE ELEVATION

PLAN

FIGURE 4-2. Orthographic projection.

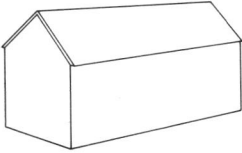

FIGURE 4-3. Perspective.

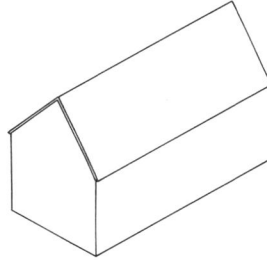

FIGURE 4-4. Isometric projection.

The customary method is to use an orthographic projection (Figure 4-2) which shows the object in plan view (from above), side elevation, and end elevation. Orthographic projections may be supplemented by sections and larger-scale details.

In order to give a clearer and more comprehensive view, other forms of projections may be used. Perspective drawings (Figure 4-3) give a foreshortened view of the whole object roughly as the eye would see it. Isometric projections (Figure 4-4) give a similar view, but they are distorted by intentionally using full dimensions without foreshortening them. Isometric projections are used more for mechanical drawings than for construction.

CONSTRUCTION TENDERS

Bids for construction projects are usually prepared and called for on behalf of the owner by the consultants. Depending on whether the project is in the public or private sector, the tender may be open or public, or there may be a limited bidders list approved by the owner. In the latter case, the number of bidders may be a function of the size of the project, except that there is generally an upper limit which is not likely to exceed ten bidders. In the case of public tenders, there is no limit on qualified contractors. Sometimes large numbers of firms take out plans, but not all are likely to submit bids for the project.

Prescribed tender forms may be used in order to make it easier to compare bids. The forms are issued to the bidders with special instructions and amendments to bid documents, if necessary. Addenda may be issued with tender documents prior to submission of tender. It is customary to stipulate that "the lowest or any bid is not necessarily accepted."

On public bids, substitutions may often void a tender, but on private projects, bidders are usually allowed to submit special qualifications as separate letters. Sometimes even these result in rejection of the bid.

Proper completion of the tender form is of extreme importance since legal implications are involved. Signatures, company seals, and resolutions may be essential and deposit checks or bid bonds, if required, must be included. Submitting the tender on time is a must. Many fully qualified low bids have been rejected because the tender was not submitted on time.

SITE INVESTIGATION

In order to assess a project properly for tendering and to construct it efficiently after the contract is received, it is important not only to study and evaluate the tender documents but also to check the site and related conditions in as much detail as necessary for the nature of the project. The larger the project, the more information may be required and the more time may be warranted to be spent on this study. It is also expedient to take color Polaroid pictures of existing conditions. The pictures will also serve as proof in case discrepancies are discovered in tender or contract documents. If extensive demolition or blasting is planned, these pictures might become very important if the adjoining property owners were to claim damages.

The checklist below enumerates some of the more important items to be recorded.

Checklist

SITE INVESTIGATION

SITE IDENTIFICATION
  Location
  Reference points
SITE ACCESS
  Good
  Bad
  Public transportation
    existing
    nonexisting
SITE CONDITIONS
  Existing buildings or structures
    To remain

    To be relocated
    To be demolished
  Slope pattern of site
    Fill required
  Precipitation pattern
    Snow
    Rain
  Climate characteristics
    Temperature
    Humidity
    Winds
  Ground water table

Checklist

SITE INVESTIGATION (continued)

Surface water pattern
   Running
   Stagnant
Vegetation pattern
   Fields
   Wooded areas
   Marshland
   Other conditions
Soil tests
   Supplied
   Required
   Surface composition
   Top soil thickness
   Subsurface composition
   Rock
   Soil-bearing capacity
   Piles or caissons required
Space available for
   Storage
   Parking
SITE UTILITIES AVAILABLE
   Sewers
      Storm
      Sanitary
   Water
   Electricity
   Gas
   Telephone
   Other
SITE RESTRICTIONS
   Rights of way

Easements
Servitudes
Homologations
Zoning restrictions
Noise restrictions
Signs
CONSTRUCTION CONDITIONS
   Labor market situation
   Local unions
   Availability of building materials
AUTHORITIES
   Building Permits Division
   Mechanical permits
   Electrical permits
   Pollution Control Division
   Highway Department
   Traffic Department
   Department of Labor
   Fire Department
   Police Department
   Power company
   Gas company
   Telephone company
CODES AND REQUIREMENTS
   Applicable building codes
   Applicable labor codes
   Climatic requirements
   Fire Department regulations
   Pollution control standards
   Noise abatement regulations
   Electrical current characteristics

ESTIMATES

During the creative cycle of a project a number of estimates are prepared. Since they serve different purposes, they vary in content, format, and degree of accuracy.

In the conceptual stage, preliminary estimates are made to assist in the preparation of feasibility studies. These estimates may start as an order of magnitude estimate to arrive at "ballpark" figures. This can be done by factoring methods, that is, by breaking up the estimate into a number of parts which can be priced on some comprehensive unit price basis, for example, cost per bed for a hospital or cost per classroom for a school, or on a more detailed basis such as per square foot (which ignores height) or per cubic foot (which takes height into consideration). Statistics from similar or comparable projects can also be used. The accuracy of such an estimate should be of the order of plus or minus 20%. If more detailed information is available, the preliminary estimates can be refined to an accuracy of plus or minus 15%.

Once the concept has been frozen and the project has been given the green light to proceed and preliminary design has been completed short of working drawings, a definitive estimate can be developed which will serve as the basis for the target cost. This estimate can be divided into different parts which can easily be separated for control purposes and will serve as global or individual control estimates for the duration of the project.

At tender stage the contractor must prepare an estimate which not only must achieve the highest accuracy but which should also be set up in such a manner that if the contractor obtains the contract, the estimate will serve as the basis of the project control system. Thus in order to be able to monitor the project effectively, the estimate must follow a factoring system that can be followed easily on the job site and coded by the supervisory site personnel. The statistics deriving from this fulfill a number of useful functions. They provide information for gauging field performance, data for preparation of progress claims, a means for the estimating department to check their accuracy and assumptions, and statistics for future projects.

Most general contractors use a similar general format for construction estimates. They may vary in details, but they usually contain the following major sections:

- Overhead items, covering indirect or nonproductive costs
- General contractor's trades, covering production costs of items that are usually priced on a unit price basis and are often performed by the general contractor on a direct-hire basis
- Subtrades, covering productive costs of items given out on subcontracts
- Profit, which can be computed in a variety of ways

A checklist of typical estimate items is given on page 159. Grouping, of course, may vary with different firms, and apart from possible overlaps, many of the items can be broken down into many subgroups according to individual preferences.

Checklist

CONSTRUCTION ESTIMATE

SITE ORGANIZATION
Personnel
   Project manager
   Project engineer
   Superintendent
   Foreman
   Instrument man
   Rodman
   Timekeeper
   Field accountant
   Clerk
   Storekeeper
   Watchman
   Equipment operator
   Mechanic
   Oiler
   Truck driver
Payroll burden
Traveling allowance
Board allowance
Temporary Services and Facilities
Permits
   Construction
   Lines and levels
   Sewer and water
   Sprinkler
   Gas
   Electricity
   Street and sidewalk
   Telephone
   Other
Bonds
   Performance
   Labor and material
   Other
Insurance
   Public liability
   Property damage
   Builders risk
   Automobile
   Plate glass
   Fire
   Other

Financing charges
Legal fees
Taxes and duties
Offices and sheds
Storage and warehousing
Trailers
Boarding and fences
Closing and partitions
Sanitary convenience
Office equipment and supplies
First aid and medical supplies
Fire protection equipment
Temporary light and power
Temporary heat and protection
Temporary water and sewer
Temporary telephone and telex
Temporary computer terminal
Temporary roads and access
Snow removal
Pumping and dewatering
Inspection and tests
Scaffolding and shoring
Trucking and cartage
Hoisting and cranes
Equipment and tools
Fuel and consumables
Small tools
Drawings and reproduction
Royalties and patent fees
Protection of work
Cutting and patching
Maintenance period service
Signs and advertising
Progress photographs
Clean glass and breakage
General clean-up
DEMOLITION　(if applicable)
Disconnecting and capping existing
utility services on construction site
   Water
   Sewer
   Gas
   Electricity

Telephone
Alarm system
Demolition of existing buildings and
    structures
Temporary closing and dustproofing
Filling in structure to grade
Removing surplus debris
Salvage of reusable materials

EXCAVATION AND BACKFILL
Removal of trees and vegetation
Protection of trees and shrubs to be
    salvaged
Stripping and stockpiling top soil for
    landscaping
Excavation
    Mass
    Trenching
    Hand trimming
    Rock
Backfill
    Excavation material
    Granular material
    Granular base under floors
    Compaction of backfill
Drainage
    Farm tile
    Interior drains
    Exterior drains and sewers
    Catch basins and manholes
Shoring
Sheet piling
Underpinning
Street cut
Street repair
Finish grading
Guardrails and curbs
Dry stone walls
Flagstone paving
CONCRETE
Formwork
Concrete
    Supply

Admixtures
Wintercharges
    Heating and protection
    Placing
Reinforcing steel
    Accessories
    Supply
    Setting
Cement finishing
Wire mesh
    Supply
    Placing
Plastic film under floors
Expansion and control joints
Saw cuts
Foundation coatings
Grouting and patching
Finishing exposed concrete
Perimeter insulation
Concrete inserts
Anchor slots for masonry
MASONRY
Masonry materials
    Brick
    Concrete block
    Terra cotta
    Glazed brick
    Glazed block
    Ornamental block
    Gypsum block
    Siporex block
    Clay tile
    Glass block
    Mortar
Masonry hardware and accessories
Masonry labor
Clean and point
Silicone treatment
CARPENTRY
Materials
    Lumber
    Plywood
    Vapor barrier

CONSTRUCTION ESTIMATE (continued)

Insulation
Building paper
Plastic laminate
Builders hardware
Labor
   Rough framing
   Millwork and paneling
   Finish hardware
   Washroom accessories
   Built-in items — concrete and masonry
   Supplied items — miscellaneous iron
   Pressed steel frames
MISCELLANEOUS SUBTRADES
   Pile foundations
   Caissons
   Structural steel and joists
   Steel decks — corrugated and cellular
   Asbestos decks
   Hollow-core slabs
   Precast concrete structure
   Prestressed concrete structure
   Lift slabs
   Laminated wood structure
   Exposed concrete aggregate work
   Asbestos siding
   Prefinished metal siding
   Cut stone and granite
   Artificial and precast stone
   Roofing and sheet metal work
   Plastic and epoxy roofs
   Skylights and monitors
   Waterproofing — membrane and metallic
   Dampproofing
   Caulking and weatherstripping
   Insulation
   Coolers and refrigerators
   Fireproofing
   Soundproofing
   Sprayed asbestos insulation
   Blown mineral wool insulation
   Sprayed cement finish
   Pressure grouting
   Miscellaneous ironwork

Ornamental ironwork
Glass and aluminum work
Curtain walls
Sliding door storefronts
Window cleaning bolts
Window cleaning machine
Steel sash
Sull sash
Fire doors
Hollow metal and kalamein doors
Pressed steel door frames
Toilet and shower stalls
Vault doors and frames
Overhead doors
Vertical lift doors
Rolling metal shutters
Rolling grilles
Rolling wood shutters
Folding doors and partitions
Millwork and wood doors
Plastic laminate faced doors
Industrial rubber doors
X-ray protection lead lining
Wire mesh partitions
Lockers
Dock bumpers
Dock levelers
Steel fences
Awnings
Lath and plaster
Dry wall work
Acoustic tile and suspension
Suspended metal ceilings
Painting and decorating
Plastic wall finishes
Furnishings and decor items
Graphics
Drapes
Blinds
Carpets
Tile, terrazzo, marble
Resilient flooring
Epoxy floors

Checklist

CONSTRUCTION ESTIMATE (continued)

Mastic floors
Hardwood floors and parquetry
Finish hardware
Washroom accessories
Asphalt paving
Landscaping
Elevators
Escalators
Moving sidewalks
Hoists
Dumbwaiters
Conveyors
Pneumatic systems
Ventilation
Refrigeration
Air conditioning

Plumbing and drainage
Heating
Sprinklers and fire protection
Supervisory and alarm systems
Electrical and substation
Music and public address systems
Chimney and incinerators
Garbage compactor
Restaurant equipment
Kitchen equipment
Theater equipment
Laundry equipment
Laboratory equipment
Gym equipment
Chalkboards and classroom equipment

CONTRACTS

A contract is defined as a legal and binding agreement between two or more persons or parties. Generally, this agreement specifies exactly what the parties are to do for each other as well as what they are not to do. To be valid, the contract must be ratified by all parties. In the case of corporations, a resolution from the board of directors may be required.

In contracts dealing with construction matters it is of special importance to treat in detail and clarify any item that could become contentious and result in delays. Since construction is time-intensive, any delay can result in loss of valuable time and in additional costs.

A contract is of a bilateral nature and it implies certain considerations by both parties, meaning that there are generally advantages accruing to each party. These advantages, however, do not necessarily carry equal weight. Furthermore, the spirit of the contract carries a certain intent which is meant to legally bridge the gaps left by limitations of language, such as the limits to which items can be described as well-defined entities. It is possible for certain intent to be implied only (based on the state of the art or good professional and trade practice) but implying intent may lead to dangerous situations if the implications are too general, are vague, or are unsubstantiable if they are challenged.

Quantitative considerations are usually easier to resolve than qualitative considerations because qualitative considerations deal with much more abstract matters. If terms such as "equivalent" or "acceptable" are introduced in contract documents, it may become difficult to establish commensurate standards and arrive at acceptable decisions.

An important, though abstract, component of contracts is good will. In practical terms, good will means that the parties cooperate in order to prevent small problems from snowballing into larger problems.

Perhaps it can be called a sign of the times, but a large proportion of contract conditions today deal with default and the measures to be taken by either party if there is a default.

There are certain clauses which can nullify contracts in whole or in part if they contain unenforceable or illegal provisions. It does not matter whether these unenforceable or illegal provisions happened by accident or by intention and they will not stand up legally even if the parties concur on the point.

A delicate item that often requires special coverage is escalation of costs, whether these are costs for labor, material, equipment, or other items. Since escalation of costs is tied in to a time factor, these costs must be adequately monitored and controlled. It is most important that the contract insist that the parties maintain completely detailed "as built" records and schedules. These records and schedules and the "as planned" schedules will permit proper evaluation at some future date.

There are primarily three types of contracts used in the construction industry. In chronological order they are:

A. Contracts with consultants
B. Contracts with general contractors
C. Contracts with subcontractors

The following sections deal with various aspects of these contracts.

CONSTRUCTION CONTRACTS

Most projects, regardless of whether they are handled through conventional construction methods or on a PM or CM basis, require a general contractor and/or subcontractors. Since the information supplied may vary considerably, it is necessary to suit the nature of the contract to the method used in implementing the project. With the growing trend toward design-build projects, that is, projects that are designed concurrently with their construction, it is often neither possible nor expedient to have a conventional all-inclusive lump-sum contract. With PM or CM, a number of bid packages may be given out piecemeal and then coordinated. The choice of contract format is also influenced by the choice of contractor. The following alternative formats could be used:

A. Lump-sum contract
B. Cost plus percentage contract
C. Cost plus fixed fee contract
D. Cost plus fixed fee plus upset price contract
E. Cost plus percentage plus upset price contract
F. Cost plus fixed fee plus upset price plus savings participation contract
G. Cost plus percentage plus upset price plus savings participation contract

The upset price limits the upper limit of costs and it is up to the contractor to stay within these limits or be responsible for any overrun. In the savings participation clause any difference in cost left at the end of the project between the total cost and fee and the upset price is divided between the contractor and the owner in a prearranged proportion stipulated in the contract.

The formats mentioned above are compared in more detail on the following pages.

A) Lump-Sum Contract
  Costs: Final or maximum, except for changes, known at start
  Cost Control: very good
  Changes: Costs may be high,
    implementation fairly hard
  Contractor's Incentive: Maximum incentive to keep costs low;
    maximum incentive to complete fast
  Design: Plans and specifications must be very tight;
    information must be as complete as possible;
    there is no flexibility
  Duration: Should be minimum since control can be good
B) Cost Plus Percentage Contract
  Costs: Known only at end of project

Cost Control: Very little or none
Changes: Costs will be high
    implementation is easy
Contractor's Incentive: None to reduce costs:
    none to complete fast
Design: Plans and specifications can be very loose;
    information can be minimal;
    there is maximum flexibility
Duration: There is no control
C) Cost Plus Fixed Fee Contract
   Costs: Known only at end of project
   Cost Control: Can be difficult
   Changes: Costs will be fair
      implementation is easy
   Contractor's Incentive: Limited to reduce costs;
      limited to complete fast
   Design: Plans and specifications can be fairly loose
      information can be limited
      flexibility is high
   Duration: There is poor control
D) Cost Plus Fixed Fee Plus Upset Price Contract
E) Cost Plus Percentage Plus Upset Price Contract
   Costs: Known at start of project except for scope changes
   Cost Control: Fair to good
   Changes: Costs will be low
      Implementation may be difficult
   Contractor's Incentive: Limited to reduce costs;
      limited to complete fast
   Design: Plans and specifications can be fairly loose;
      information can be limited;
      flexibility is fair
   Duration: Control is fair
F) Cost Plus Fixed Fee Plus Upset Price Plus Savings
   Participation
G) Cost Plus Percentage Plus Upset Price Plus Savings
   Participation
   Costs: Known at start of project except for scope changes
   Cost Control: Very good
   Changes: Costs will be low;
      implementation may be difficult
   Contractor's Incentive: Very high to reduce costs;
      very high to complete fast
   Design: Plans and specifications can be fairly loose;
      information can be limited;
      flexibility is fair
   Duration: Control is fair

DEFINITION OF COSTS

It is essential to define costs in sufficient detail and indicate which costs are reimbursable under the terms of the contract. Overlaps and ambiguities must be avoided. Site costs generally do not present any special problems, but head office costs may create difficulties. In particular, it is important to differentiate between personnel belonging to the project or site organization in whole or in part but operating out of the head office for convenience in view of their interaction with other personnel and departments (accounting, payroll, etc.) and the staff who belongs to the head office but may make an occasional site visit, which might be classified under head office supervision and may not be a legitimate expense since it may be included in the fee. Head office costs can be expressed in different ways. They can be included in the percentage fee, as mentioned above, applied to the cost of the project, they can be taken as a weekly or monthly allowance, they can be calculated as a percentage of site costs, they can be equated to labor costs, or they can be handled on any other mutually acceptable basis.

In listing costs, special attention must be given to trade and payment discounts.

Trade discounts, which may be given on large material purchases, on orders involving duplication of items, or for other reasons, are credited to the owner.

Payment discounts are given for payment within stipulated time limits and should be credited to the party responsible for payment.

The contract should cover the following items under reimbursable costs. Other items may be added as relevant.

Checklist
REIMBURSABLE COSTS

Labor
> Salaries and wages for all personnel for time actually
> spent on project, whether on site or elsewhere, with
> other locations clearly spelled out in contract.
> Head office and other locations: As defined under terms
> of contract. Personnel might include: project and/or
> construction manager, administrative assistants,
> engineering personnel, estimators, planners, cost
> controllers, schedulers, expediters, procurement personnel,
> purchasing agents, accounting personnel, clerical staff.
> Site office: site manager, superintendents, foremen,
> field engineers, surveyors, rodmen, field accountants,

clerks, warehouse man, safety personnel, first aid staff,
inspectors, tradesmen, etc. Payroll burden for all personnel and all other applicable charges, such as unemploy-
ment insurance, workmen's compensation, medical insurance,
vacation pay, traveling, various other applicable benefits,
out-of-pocket expenses.

Company Overhead

Materials and Equipment

All permanent materials and equipment and other supplies
bought for the project including all applicable sales
taxes, freight charges, duties, cartage, demurrage,
loading, unloading, handling, brokerage fees, exchange,
insurance.

Rentals

All valid rental charges for equipment used on project
including taxes, cartage, handling, assembly and dismantling,
repairs and maintenance, other applicable charges.

Subcontracts

All subcontracts awarded on behalf of project plus all
valid adjustments and changes.

Temporary Services and Facilities

Construction or rental including maintenance and repairs,
dismantling and disposal of all temporary offices,
warehouses, sheds, and other structures, hoarding,
enclosures, barriers, etc. Installation and removal of
temporary services for water, drainage, sanitary
services, electricity, telephone, security, heating,
ventilation, air conditioning, fire protection, safety,
first aid, P.A. system, etc. Temporary equipment,
small tools, consumables, stationery, other supplies,
office equipment, etc.

Miscellaneous

Permits, licenses, royalty payments, insurance and bond
premiums, any other contractual expenses, etc.

Trade and Payment Discounts

Clarification of terms for this item.

Items Supplied to Contractor for Installation

Terms to be clarified for this item.

GENERAL AND SPECIAL CONTRACT CONDITIONS AND CLAUSES

Date of contract
Identification of parties
Identification of project
Scope of project
Scope of contract

Contract completion date
Definitions of terms
Obligations of contractor
   To perform the work assiduously
   To supply sufficient and competent personnel
   To provide insurance and bonds as agreed
   To make only authorized commitments
     on owner's behalf
   To provide office facilities for owner's
     representatives and consultants
   To keep complete and detailed records of all
     transactions concerning project and to
     afford access to same on request
   To comply with all applicable laws
   To cooperate closely with owner's
     representatives and consultants
   To submit within agreed on time limit project
     schedules and organization charts required
     for project monitoring and control
   To submit progress claims at times and in
     format as agreed
   To hold the owner harmless from any of its actions
     or that of its subcontractors
   To assist the owner in dealing with the
     authorities concerning the project
   To use only new materials except when
     specified otherwise
Owner's obligations
   To provide an official representative authorized
     to deal with contractor and make prompt
     decisions as required
   To furnish contractor promptly with all pertinent
     data and information required for effective
     implementation of project
   To approve drawings and other documents
     expeditiously in order to avoid delays
   To obtain and pay for permits required for the
     project
   To remunerate the contractor in terms of the
     contract for services rendered and work
     performed
   To pay for the entire cost of the work in terms
     of the contract
Changes
   Basis of extras and credits

Implementation
Plans and specifications
Number of sets issued
Ownership
Insurance coverage
Builder's risk
Liability
Property damage
Fire
Automobile
Sprinkler
Other as agreed
Guarantees
Determination of substantial and final
completion
Subcontractors' guarantees
Separate contracts and subcontracts
Inspection
Rejection and replacement of unacceptable
materials and workmanship
Termination, suspension, delay, abandonment
Conditions for settlement for initiation
of any of above by either party
Default by either party
Force majeure
Arbitration
Headings of articles in contract
Waivers of rights or privileges
Successors and assigns
Language of contract
System of measurements used
Basis of law of contract
Notices and communications
Contract amount
Lump-sum, open, upset price
Remuneration
Fee structure
Lump-sum, cost plus, fixed fee,
savings participation
Payment/holdback conditions
Release of final holdback
Treatment of changes
Definition of costs
Progress claims procedure
Signatures of all parties

EXTRAS AND CREDITS

Most completed projects are somewhat different from the original concept. During construction a certain number of changes are made, resulting in various extras and credits. If there is no major change in scope, the percentage of these changes in terms of the original contract amount gives a measure of the control exerted over design and construction. Anything up to 5% may be considered exceptional, anything up to 10% may be considered acceptable, anything between 10% and 20% indicates that there has been some laxity, and anything over 20% indicates poor performance.

In order to ensure a minimum of changes, the design must be very tight, and it must be realistically tied into existing conditions. In addition the terms must be well defined in the contract.

There are various ways of dealing with extras:

A. They can be negotiated.

B. They can be handled on a net cost of labor, materials, equipment, etc. basis plus a percentage fee for overhead and profit. The percentage varies for the general contractor and subtrades, and a special coordination allowance is made for the general contractor to coordinate the subtrades.

C. There can be any other mutually acceptable method.

When claims are submitted, they must be accompanied by complete documentation to permit substantiation.

It is customary to allow a fee for extras and deduct credits at net cost because the fee for the credit portion is already in the contractor's bid price. Strangely enough, it is possible for consultants to lose money on processing credits because it cuts down on their fees, as illustrated in the following example.

EXAMPLE OF EFFECT OF EXTRAS AND CREDITS ON CONSULTANTS

Assume that a contractor gets an $8,000,000 contract. Extras are added with a 10% fee; credits are deducted at net cost. The consultants receive a 5% fee.

Case 1: There are no extras or credits
Contractor receives	$8,000,000
Consultants receive	400,000

Case 2: There are $400,000 in extras

Contractor receives	$8,000,000
	400,000
	40,000
	$8,440,000
Consultants receive	$ 422,000

Case 3: There are credits of $400,000

Contractor receives	$8,000,000
	−400,000
	$7,600,000
Consultants receive	$ 380,000

Case 4: There are $400,000 in extras
and $400,000 in credits

Contractor receives	$8000,000
	400,000
	−400,000
	$8,000,000
Consultants receive	400,000

If the processing of extras and credits is considered as entailing the same amount of work, it can be seen that the consultants are short-changed in cases 3 and 4 by $40,000 in each case.

PROVISIONAL CONTRACT (P.C.) ALLOWANCES

It often happens that certain parts of a building cannot be designed or established in sufficient detail to permit a contractor to bid on them. In such a case, it is customary to include the items as provisional contract (P.C.) allowances in the tender documents and to adjust them at a later date against actual costs. It is important to clarify the conditions of this adjustment; for example, the tender documents may state that any fee for an item is to be included by the contractor in his basic bid and the P.C. sum will be adjusted for the net difference between the amount included and the actual cost plus taxes, duties, freight, handling, etc. If the actual cost exceeds the P.C. sum by say more than 15%, the contractor may be entitled to additional fees for the difference. Conversely, if the P.C. sum is greater than the actual cost, etc., then the contractor must pass on the net credit to the owner.

P.C. allowances should not be confused with the unit prices that are sometimes included in contracts and often are meant for conditions which could possibly be encountered but are not actually known to exist, for example, rock excavation which may or may not be required and which may require different unit prices, depending on the nature of the rock and the method of its excavation and disposal.

Items which could become subject to P.C. allowances are given in the checklist below. The items may vary on different projects, depending on the circumstances and the consultants.

Checklist

PROVISIONAL CONTRACT (P.C.) ALLOWANCES

Pile Foundations	Finish hardware
Structural steel	Electric lighting fixtures
Curtain wall	Miscellaneous equipment
Metal windows	Gymnasium
Storefronts	Laboratory
Office Partitions	Kitchen
Laboratory and field tests	Cafeteria
Soil	Classroom
Concrete	Laundry
Steel	Art work and decor
Roof	Miscellaneous items that cannot be
Miscellaneous materials	finalized at time of tender
Face brick	

BUILDING TRADES GUARANTEES

A number of trades must provide certain guarantees for their work. These guarantees are integral conditions of their subcontracts. They may be based on local codes, be set up by the consultants, or follow customary trade practice.

Among more common guarantees are the following:

Guaranteed Item	*Guarantee Period (Years)*
Paving work (bituminous)	2
Precast concrete	5

Guaranteed Item	Guarantee Period (Years) (continued)
Aluminum windows	5
Curtain wall	5
Roofing and sheet metal	2
Plastic laminate work	2
Waterproofing (membrane type)	5
Glazing and mirrors	5

It should be noted that the above are supplemented and may be overruled by legal warranties as stated in existing applicable legislation.

PLANNING AND SCHEDULING

In order to meet specified completion dates and keep project duration to a minimum, there must be proper advance planning. Once a plan has been prepared, schedules are drawn up. These schedules are used to aid in the implementation of the planning and for control purposes.

If a project is to be planned effectively, the project must be broken down into various definable and measurable operations which should be separated according to trades. Although it is advisable to have a well-detailed plan of action, discretion must be applied in the type of operation noted and the degree of detail pursued. The plan of action should be flexible so operations can be changed if necessary.

On most projects the general contractor prepares a project schedule. In fact, this schedule often constitutes an integral condition of the contract. If projects are handled through PM or CM, the owner or consultants prepare a suitable schedule.

Over the years, different methods have been used to implement scheduling and these methods have been refined continuously up to the level of present-day network techniques. Not only are network techniques extremely sophisticated but because they are based on mathematical notation they have introduced the application of computer techniques and programming to scheduling in all phases. Network techniques permit not only the planning of the project but also continuous monitoring. They indicate whether the project is on time, they show where slippage, if any, is occurring, and they sound the alarm on potential delay problems and bottlenecks.

The basic and simplest means of scheduling is the bar chart, which, despite its shortcomings, still finds extensive use and popularity, especially with people who have trouble understanding the principles of network techniques.

The bar chart (Figure 4–5) breaks up the project into a number of operations on a time scale. It shows the beginning and end of each operation. The problem is that the bar chart cannot show the true relation and interface with other operations and it does not show if the project is on time, ahead, or behind.

Attempts have been made to improve the bar chart. Among a number of the variations of the chart is the well-known Gantt chart (Figure 4–6). The Gantt chart has a combined parallel "as-built" chart tied into the time scale to indicate the relationship of a particular operation to its theoretical duration. What distinguishes it is that the heavy line drawn underneath represents the amount of work in proportion to the total work of the operation. A special mark indicates the revision date of the chart and the state of progress of the operations. In the example shown, Operation 1 is on time, Operation 2 is behind, and Operation 3 is ahead.

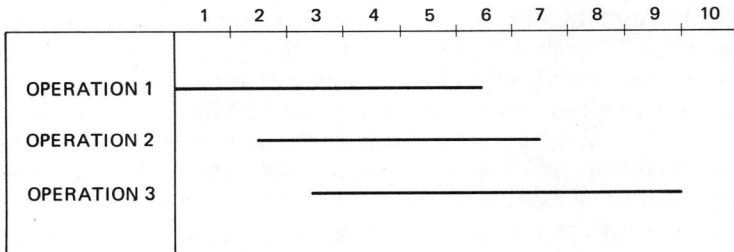

FIGURE 4–5. Bar chart.

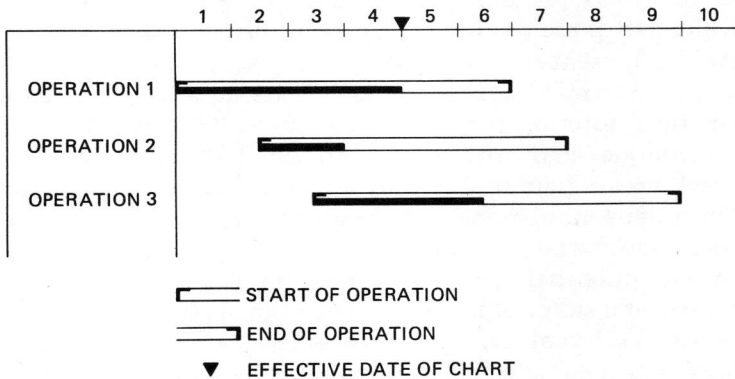

┌──────┐ START OF OPERATION

───────┐ END OF OPERATION

▼   EFFECTIVE DATE OF CHART

FIGURE 4–6. Gantt chart.

Bar charts have been improved by the addition of milestones. These are events that have definite cut-off points in time and can, therefore, be easily defined.

The main problem with the Gantt chart and similar charts is that they are based on a time scale and are dependent on adherence to that scale. If there is any deviation in one activity, the relationship to the other activities is changed in a way which cannot be determined from a bar chart.

The changes which revolutionized the industry in this respect came in the 1950s when network analysis schedules came into being. The U.S. Army developed the program evaluation and review technique (PERT) (Figure 4-7) to control the extremely complex Polaris Missile program.

The DuPont company developed the critical path method (CPM) (Figure 4-8) for use in the construction of complicated chemical plants.

From these systems other methods evolved, for example, the precedence diagram (Figure 4-9), with other modifications and improvements appearing occasionally.

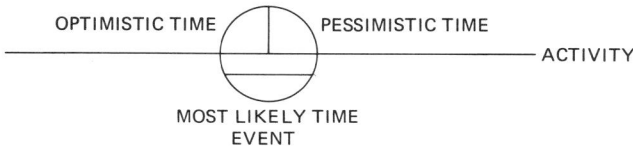

FIGURE 4-7. PERT.

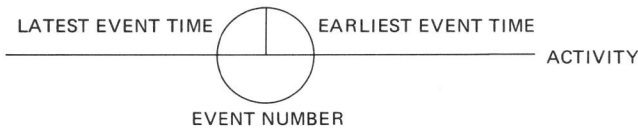

FIGURE 4-8. CPM.

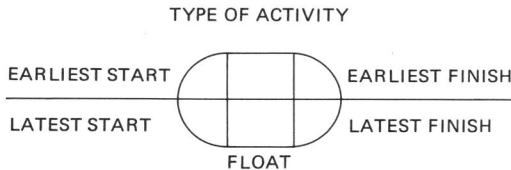

FIGURE 4-9. Precedence diagram.

All these methods have certain features in common, including the most important one of showing individual interrelationships among the various operations with respect to time. Since these are based on numerical terms, they are easily computerized. They permit early prediction of future requirements, which in turn simplifies ordering and expediting of equipment and materials and scheduling of labor. They also show when certain operations are falling behind. They permit the exercise of manpower leveling, an extremely important activity which aims for the optimum use of the available labor force. On projects at distant locations, where manpower is not flexible and probably at a premium, this is an extremely important consideration.

A few of the more important network techniques are described briefly below.

PERT

This is a technique in which events, shown as nodes on the network, are connected by related activities, shown as arrows. This method introduces probability by using both the most likely and least likely time factors. It is hoped that the activities are completed within these time factors.

CPM

The symbols are similar to PERT in this method, but the activities are related to each other in a way that presupposes a reasonable time for execution. Any deviation from this time factor represents a slippage, which is seen as a loss of control. Activities that take less time than allotted to them in terms of parallel activities are considered to have float time. This float time can be used at times, depending on the circumstances, for corrective measures. The critical path is obtained by joining activities that have no float between the beginning and end of the project.

Precedence Diagram

This is similar in principle to the other methods, but the symbols have been reversed. The interconnected nodes represent activities and show both most and least likely start and finish of these activities and indicate possible float time. Thus the events are virtually tied in on a variable basis.

In retrospect, it can be said that each method has certain advantages and disadvantages, but they all have one thing in common. They are control guides only to be used in an auxiliary capacity. The fact that a project has a network diagram does not guarantee that the project will be completed on time. Actual control must be judiciously exercised by using the information supplied by the network. The network can indicate when there is danger of slippage before it actually occurs, but from there on the action is up to project management.

Checklist

MILESTONES

Note: In order to prepare a meaningful schedule and maintain good job records and control, certain events that can be pinned down to close dates should be singled out. The following list indicates some examples:

Contract award
Job mobilization
Starting/completion
   General excavation
   Pile foundations
   Footings
   Foundation walls
   Structure to grade
   Structure to roof (by floors)
   Basement slab
   Roofing
   Curtain wall and windows (by floors)
   (building closed in)
Start/complete roughing-in (by floors)
   Interior partitions
   Doors and frames
   Ceiling suspension
   Water and drainage
   Sprinkler system
   Heating pipes
   Ductwork
   Electric conduits
   Telephone
Installation
   Boilers and furnaces

Air handling units
Fan coil units
Radiators and convectors
Fans
Condensers
Cooling towers
Bus ducts
Panels
Substation and transformers
Electric wiring and panels
Telephone wiring and panels
Finishes (by floors)
   Wall finish
   Doors and hardware
   Ceilings
   Flooring
   Sprinkler heads
   Plumbing fixtures
   Grills and diffusers
   Lighting fixtures
Clean-up
Acceptance
Handing over building to owner

INSPECTION AND TESTS

Various types of work require inspection and testing services by competent laboratories and inspection firms. Although the scope of these inspections and tests vary from one project to another, it may be convenient to list some of the more common items to be covered under this subject:

Earthwork
  Soil bearing values
  Soil composition and density
  Suitability of soil as fill
  Compaction of subsoil
  Placing and compaction of fill
  Grading and compaction of granular material
  Bore holes and core drills
  Composition of rock
  Rock bearing values
  Water table
Concrete work
  Cement (strength, quality)
  Aggregate (strength, composition)
  Water (purity, hardness)
  Admixtures (quality, proportion)
  Concrete mix (strength, test cylinders, slump,
      proportion, placing, vibrating, curing,
      finishing, protection)
Reinforcing and structural steel
  Alloy composition
  Fabrication (bending, forming, welding)
  Erection (setting, welding, riveting, bolting,
      guying, plumbing, painting)
Precast and prestressed concrete
  Manufacturing
  Anchorage
  Reinforcing (rebar, tendons, anchors)
  Concrete strength
Roofing and sheet metal
  Materials (weight, thickness, plies, gauge)
  Workmanship
Membrane waterproofing
  Materials (weight, thickness, plies)
  Workmanship
Aluminum work
  Material (finish, gauge)
Fire doors and dampers
  Fire rating
Paving
  Materials (composition, weight)
  Application (thickness, layers, compaction)
Other materials:
  According to project requirements

BONDS

Contractors are often required to submit bonds on projects to guarantee their work. The contractors may have to include a bid bond with their tender and, if successful in obtaining the contract, must supply the bonds as stipulated. The premium may either be paid by the owner or be included as part of the contract amount.

The following bonds may be asked for:

- A 100% performance bond
- A 50% performance bond and a 50% labor and material bond.

These bonds represent the most frequent bonding requirements for general contractors, but other variations are possible.

It must be emphasized that having a bonded contractor does not necessarily guarantee a successful project. The bond does, however, give some financial protection to the owner against failure or default by the contractor, usually for unexpected reasons, since the bond, depending on its nature, is usually designed to cover the cost of completing any outstanding work on the project.

The premium paid for the bond is usually for a one-year period from the date of contract signature. At the end of the first year the bond is renewable based on the balance of contract to complete, with the premium prorated accordingly. The same conditions apply on a multi-year contract until completion.

INSURANCE

During construction certain insurance coverage is required. This insurance is very different from the permanent insurance taken on the building after it has been accepted by the owner. This coverage may be of a variable nature and the responsibility is gradually transferred from the contractor to the owner as completion advances. It is important to specify this gradation clearly in the contract as well as in the policies so that if there is a claim there is no question about who is responsible for the insurance and what the limits of coverage are. Initially the contractor may be responsible for full insurance coverage of the project. As the project continues, the owner pays for it in progress claims, often taking out insurance to cover the paid for part, while at the same time the contractor's liability is reduced. There is not necessarily a special stage at which this happens, rather it is a gradual process.

Insurance coverage varies on different projects. Some types of insurance are listed below:

- Builders risk
- Public liability
- Property damage
- Fire insurance
- Sprinkler insurance
- Motor vehicles insurance
- Plate glass insurance
- Elevator insurance

TEMPORARY FINANCING

When general contractors prepare tenders, they must make a certain allowance for temporary financing during construction.

Progress payment applications may be prepared for work done during the previous month and submitted on the 15th of the current month. These applications are approved and paid only by the end of the month or later, depending on contractual terms. This means that the contractor has from one to two months time lag before getting paid. He may be able to defer payments to subcontractors and suppliers, but he must pay his direct labor and other expenses at once. In addition, he must make allowance for the holdback deduction from his progress payments which may amount to as much as 15% and is retained until some time after completion and acceptance of the project. All this he must finance temporarily and the interest can amount to a substantial amount. There are firms that specialize in financing contractors during the construction period.

Financing the overall operations of the contractor is a different matter altogether. In this case, the contractor has to rely on the good will of his bank and other financial institutions which, in turn, depends on the contractor's track record.

SUBSTANTIAL COMPLETION

When a building approaches completion, it is at a stage where it is not entirely complete but, apart from the correction of deficiencies, only very little work remains to be done. Actual completion of the project, however, can drag out for a long time. If at this point the building is usable as intended, it can be said that substantial completion has been reached. The following criteria can be used in this respect.

In the absence of progress statistics, progress claims could be used. A percentage of about 97½ % is suggested as a reasonable basis for establishing substantial completion.

In terms of actual building construction components, the following items should be essentially complete:

- Building structure and frame
- Exterior walls
- Interior partitions
- Mechanical systems (heating in winter, air conditioning in summer)
- Electrical systems and permanent power
- Most interior finishes (floors, walls, ceilings)
- Most rental area work (tenant services)
- Parking and site services

Items left to be completed are primarily of a nonessential nature and do not substantially interfere with use of the building.

Substantial completion can be certified by the consultants on the owner's behalf or acknowledged by the owner acting as his or her own project manager. Substantial completion does not mean final acceptance. Final acceptance can only be accorded after all outstanding work is completed and all remaining deficiencies are corrected. It is, however, possible, and fair to the general contractor, to release holdback money, retaining enough to ensure coverage of items still to be completed or corrected. At the same time, it should be made clear that this action is taken without in any way, and without any prejudice to the owner, relinquishing any rights of the owner toward the general contractor regarding guarantee periods, legal rights, or any other related aspects. The owner may release some holdback money to the contractor. Without prejudice means that by doing so the owner does not accept the project or rescind or abrogate any rights he or she may have with regard to the contract with the contractor. In any case, the latter must complete the contract. Receiving a partial release of holdback money is only intended to ease the financial burden. Since holdback is generally 10% of the progress payments, the contractor may work on a smaller profit margin.

HANDING OVER PROJECT

When the project is completed and the contractor hands the project over to the owner, the contractor supplies the owner with the following:

- Keys (unless all cylinders are going to be changed). It is possible to equip the hardware installed during construction with temporary inserts and temporary keys. When the building is handed over, the inserts are removed and the permanent keys are given to the owner who does not have to change the cylinders.
- "As-built" drawings. These drawings are based on construction drawings and show all changes made after the original contract drawings were made and changes made during construction. They also indicate the location of piping, conduits, etc. and the location of all building elements required in the operation and maintenance, such as valves and switches, all of which must be clearly shown and tagged.
- Manuals. These manuals are for the different systems in the building and they describe the operation and the maintenance of these systems. Included in the manuals are detailed instructions and manufacturer's data.
- Certificates of warranties and guarantees, bonds, insurance policies, and other items specified in the contract.
- Certificates of inspection and acceptance by various authorities as well as by consultants and mortgage company.
- Spare parts and tools for equipment as specified.
- Extra materials for repairs as specified.

PROJECT ORGANIZATION

The organization of a project depends on the size of the project, whether it is owner-built or contracted out, whether or not it is a design-build project, and whether or not it is handled by an independent construction manager.

In project or construction management there is a certain interaction among the several parties involved in the project, for example, the owner and his organization, the consultants and their organization, and the project itself, which may be represented by a general contractor or have its own organization. Typical organization charts are given in Figures 4–10 through 4–12. There are many variations possible and the specific requirements of each project must be considered separately.

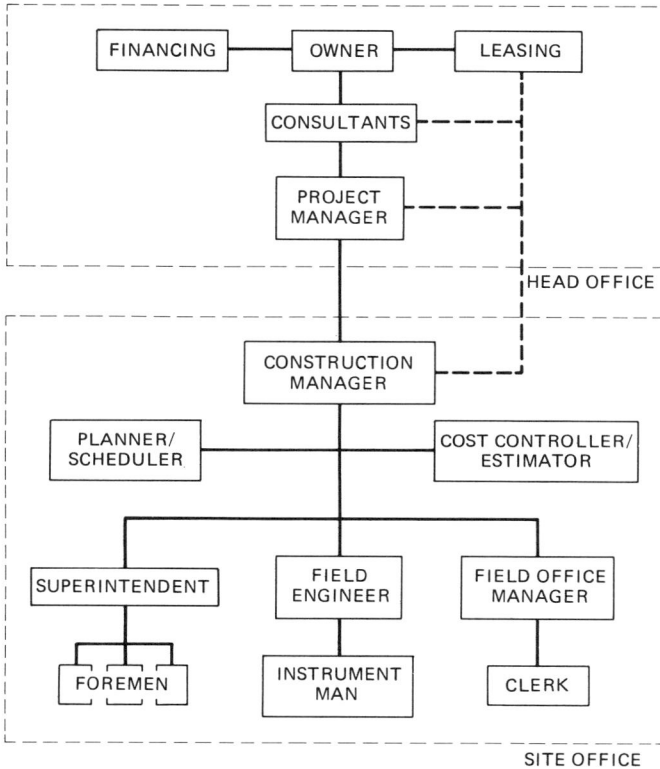

FIGURE 4-10. Example of organization chart for small project.

PERSONNEL

The success of a project depends to a great degree on the people who manage and administer the project. The secret of this success lies in organizing the right number of people in the organization structure and ensuring that all of these people carry out their functions to the best of their ability. The best way to do this is to write a complete job description for each position on the organization chart and to structure the hierarchy so that it does not create administrative conflicts. Job descriptions make the task of looking for people and filling positions much easier. They should be as specific as possible and they should delineate the limits of the jobs so that there will be no conflict later about who does what. The checklist on page 184 is a sample job description within the framework of the organization chart.

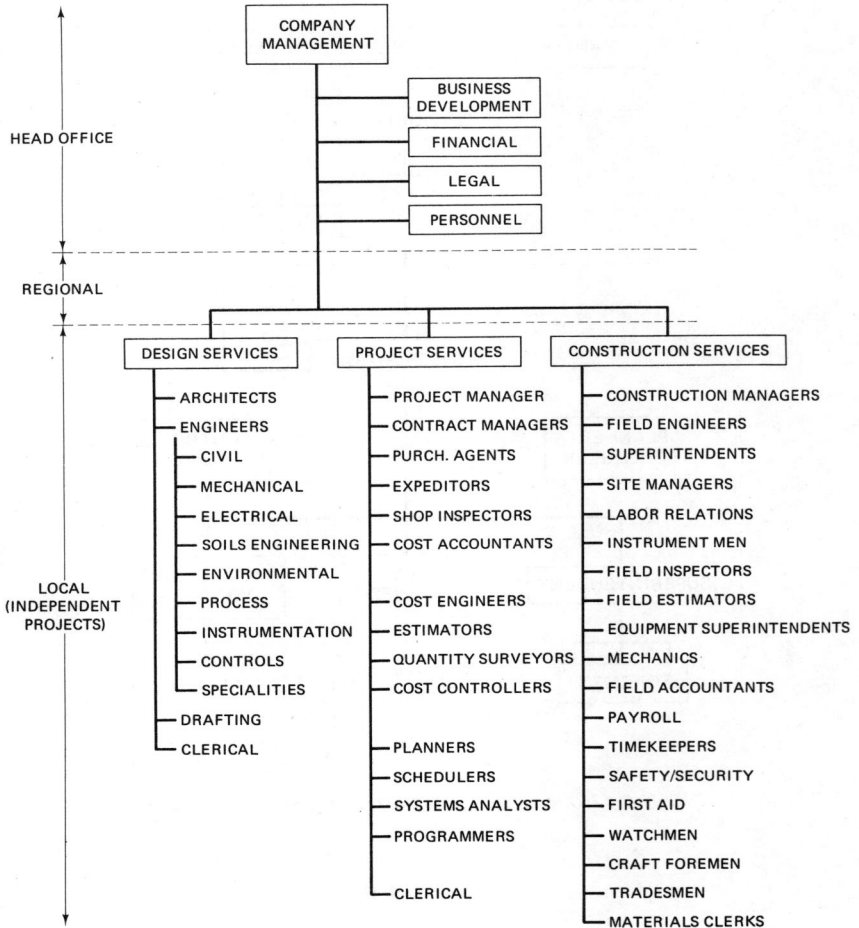

FIGURE 4-11. Example of organization chart for large design/build company.

Checklist

JOB DESCRIPTION

Job Title: (Name of position on organization chart)
Objectives of position: (Broad outline of main functions)
Responsible to: (Immediate superior, format and extent
   of reporting)
Range of Responsibilities: (Specific tasks covered
   under scope of position)

PROJECT MANAGER

DESIGN GROUP	SERVICES GROUP	CONSTRUCTION GROUP
ARCHITECTURAL	PROCUREMENT	CONSTRUCTION MANAGER
CIVIL	CONTRACTING	SUPERVISION
MECHANICAL	PURCHASING	FIELD ENGINEERING
ELECTRICAL	EXPEDITING	SURVEY TEAM
SOILS ENGINEERING	CHECKING	EQUIPMENT
ENVIRONMENTAL	PLANNING	MATERIALS
PIPING	SCHEDULING	STORAGE
INSTRUMENTATION	EDP SYSTEMS	LABOR RELATIONS
CONTROLS	COST ENGINEERING	SAFETY/SECURITY
MATERIALS HANDLING	ESTIMATING	PAYROLL
TRAFFIC	COST ACCOUNTING	CLERICAL
SPECIALITIES	COST CONTROLS	
CLERICAL	CLERICAL	

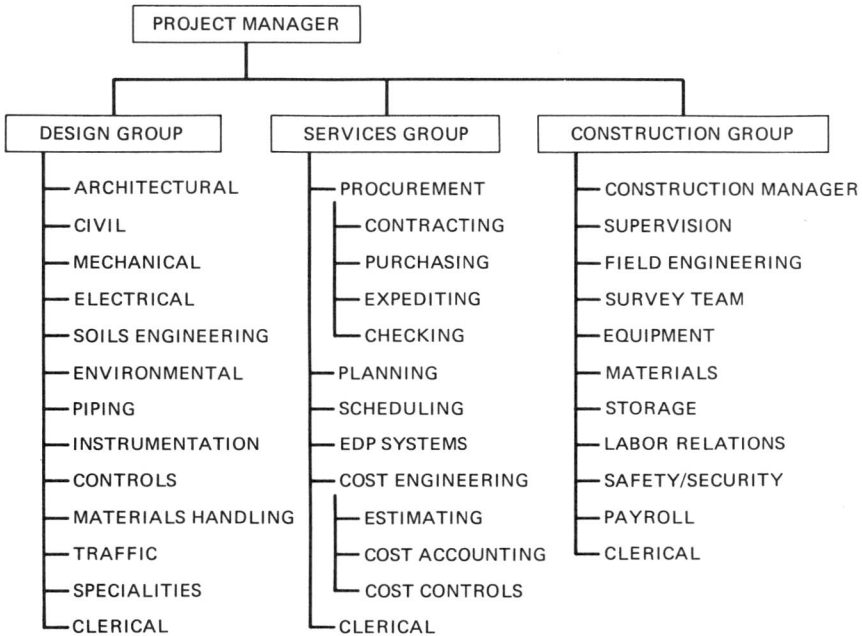

FIGURE 4-12. Example of project organization chart for large project—design/build.

Limits of Authority: (Commitments inside and outside
    company, including financial transactions, control over others
    within company)
Detailed corollary of functions: (Itemized list of duties
    and tasks)
Interrelationships: (Relationships in company with persons
    other than directly above or below)
Performance Standards: (Criteria for judging person
    filling position)
Job Profile: (Theoretical description of person to fill
    position, including age range, educational background,
    professional experience, personality type, etc.)

LABOR RELATIONS

In many industries the human element has always been the last to
receive recognition or consideration. Construction has been no exception. In an age in which work environment fringe benefits have in-

creasingly taken on more importance it is easily understandable that construction men who by the nature of their work must put up with a lot of inconvenience have become much more concerned about their working conditions. Labor unions are continuously demanding more fringe benefits over and above pay increases. The sad fact is that productivity is not keeping up with the   paycheck. Pride of workmanship went down the drain a long time ago.

The main objective of most construction companies is to maintain peace on the job site and avoid any friction that could lead to costly labor problems. In order to have a good work climate on the site, it is important to create good morale and a positive attitude in the work force. This encompasses among other measures:

- Creating interest in the project and a feeling of personal involvement and accomplishment among the men
- Ensuring good job security and accident prevention
- Providing acceptable ancillary services (canteen services, sanitary facilities, lunchroom and locker rooms, first aid, etc.)
- In the case of construction camps on remote sites, providing good recreational facilities and providing good quality food in the mess hall are musts.

Often trivial matters can take on enormous proportions and create havoc, thus triggering wildcat strikes.

Although the project manager may be the highest ranking authority in the project hierarchy, it is usually the superintendent who more than anybody else controls the site and holds the men together. Not only must he have a strong personality, he must also know his work well if he is to earn the respect and confidence of his men. In addition, he has to be able to deal with the union stewards who often tend to be rather militant.

In order to induce people to do their best, people must be motivated. Improved working conditions or higher pay has not usually been sufficient to supply this motivation in construction. It has been found that arousing the men's interest in what they are doing has proved to be much more effective. By informing the work force regularly about the project and its progress it is sometimes possible to generate a feeling of personal ties, which, in turn, results in increased productivity. This is not only important in terms of job efficiency but also in terms of job safety. A project on which the morale is good usually shows a better safety record than one on which the men are not motivated. Curiously enough, it is much easier to generate interest on projects which are not too large. On very large projects, those that last several years and employ a great number of men, there is usually a large turnover of personnel. Relationships among members of the work force become much more casual and impersonal. In addition, the physical size of what is being built does not give the men a sense of personal achievement. Instead, it makes them feel that their contributions are an insignificant part of the whole project. This large turnover of the work force on large projects has a built-in efficiency loss factor, for every new worker must go through a learning period in which his productivity is low.

It goes without saying that productivity can make or break a project financially and timewise. This aspect and other related aspects are discussed in the following section.

MANPOWER AND PRODUCTIVITY

Every project must follow a pattern. The work force has to be built up at the beginning of the project and reduced as the project winds up, with a skeleton crew finishing up deficiencies and providing the final touches before takeover by the owner. In between these stages the objective is to keep a steady work force with minimum turnover and maximum utilization of manpower. This is particularly important on projects at remote sites where it is difficult to obtain and schedule manpower (Figure 4-13).

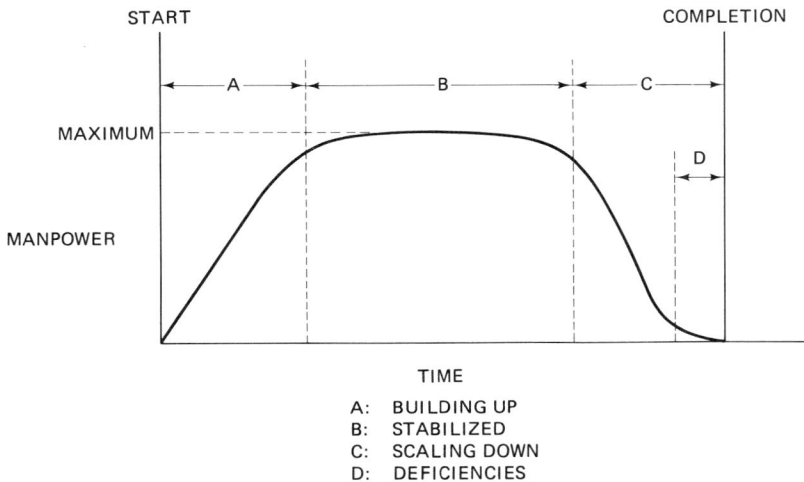

TIME

A: BUILDING UP
B: STABILIZED
C: SCALING DOWN
D: DEFICIENCIES

FIGURE 4-13. Manpower.

Productivity which, as mentioned before, is closely linked to motivation generally follows a pattern that graphically looks somewhat similar, but is phased differently (Figure 4–14). Productivity charts look similar to manpower charts, but they are based on different criteria.

As the project starts up productivity is low, but it increases as the men become familiar with the project and their own part in it. At this stage, it is important to make sure that each worker has a clear-cut picture of his own tasks and responsibilities. The men should be well informed about the planned objectives and the scheduled target dates. A simple bar chart showing actual against planned progress often helps in motivating a crew handling a specific operation.

After the learning period is over, productivity, at least in theory, should remain high until the project reaches near completion. At this point, productivity may slip for a number of reasons. The men may try to drag out the work, especially if they do not know where their next project is going to be. They tend to lose interest because the main work is completed. Correcting deficiencies and/or making finishing touches can be tedious, nitpicking, and annoying work which is never very popular. It is also more difficult to get the sub-trades back on the site for small work items.

A prime cause of loss of productivity are design changes which can sometimes be extremely disruptive. Design changes can result in holding up the project in whole or in part and losing momentum, in having to work in fits and starts, in having to perform work out of logical or planned sequence, and in having to change trades or deploy

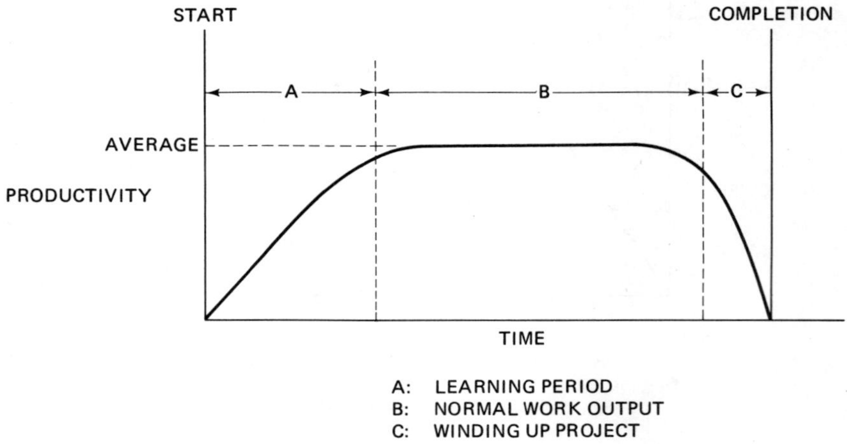

A: LEARNING PERIOD
B: NORMAL WORK OUTPUT
C: WINDING UP PROJECT

FIGURE 4-14. Productivity.

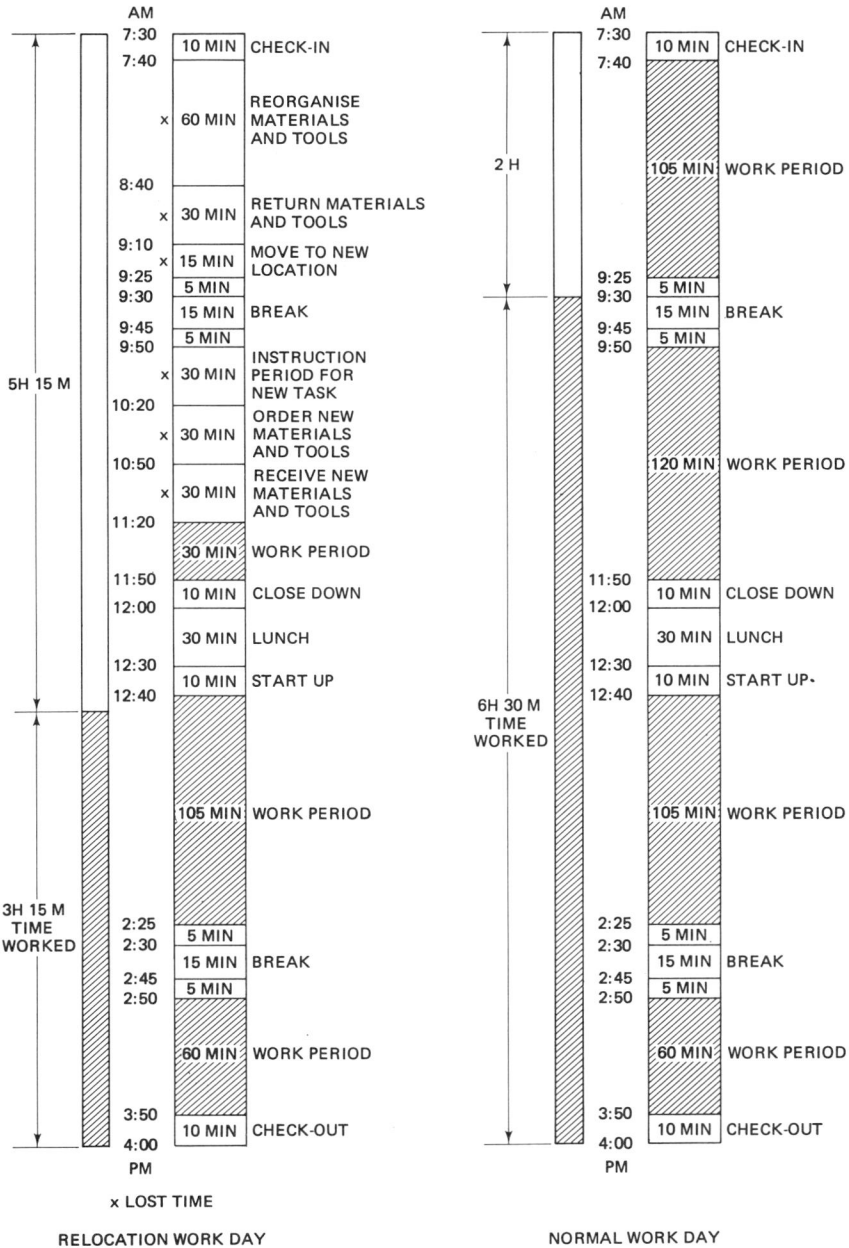

FIGURE 4-15. Working time distribution.

189

men elsewhere to mark time. Design changes may often result in claims which may be difficult to substantiate and even more difficult to quantify unless excellent and relevant records are kept (a sin of omission of many contractors).

The effect of having to relocate men on the jobsite because of changes which disrupt the work routine is shown in Figure 4-15. In this example the productive time is reduced by 50%. This means a loss of productivity will be experienced by the contractor, since the manpower is partly used on unproductive labor.

Another cause for loss of productivity is found in excessive overtime. Overtime may be worked in order to try and make up for lost time, or to accelerate a project or a specific operation. It has been found, however, that as the ratio of overtime to straight time increases, the efficiency decreases in an almost geometric proportion. A number of companies have conducted extensive studies which have shown that there is an optimum overtime increment beyond which the efficiency, and consequently the productivity, of the workforce decreases rapidly. This is not necessarily constant for all workmen, but may vary for different trades and be affected at the same time by other factors, such as weather, among others.

Temperature and climatic conditions also have an effect on productivity. Below a certain temperature (within a few degrees above freezing) and above a certain temperature, productivity decreases in proportion to the decrease or increase in temperature. Although some attempts have been made to quantify this (for example, Richardson Engineering suggests 1% drop of productivity per $1°F$ below $40°F$), any attempt to quantify this loss of productivity is an approximation at best and there is no true absolute norm to measure it by. For the average project, it can probably be gauged by comparing estimated against actual outputs, either for parts of the project or on a global basis. This presupposes that the estimate was correct and it ignores any built-in error. Units used can be either in terms of quantity, cost, or time but if the choice is not consistent, the results are meaningless.

CONSTRUCTION CAMP

Major projects on remote sites require construction camps which provide room and board as well as recreational facilities for the building trades. The camp may be supplied by the owner or it may be part of the general contractor's responsibility. Usually, the buildings are supplied directly (supplied by the contractor or owner rather than become a sub of the contractor) and the catering and housekeeping arrangements are subcontracted out to firms specializing in this type of operation. A ticket system is used for meals. The men are given special meal tickets by their firms and the catering service is paid according to the number of tickets collected. These firms also supply the bedding and furniture for the rooms and equipment for the recreation building. They may also set up a canteen service or handy store for miscellaneous dry goods. Box lunches may be supplied when it is too far for the men to return to the mess hall for lunch.

The quality of the camp has an effect on the morale of the men on the job site because the camp often provides the only comfort for the men during long periods of isolation from the outside world. Whenever possible, TV is provided (sometimes it requires special installations) and movies are shown at regular intervals. Although alcohol may be dispensed, there should be strict control over it to ensure that it does not create problems that are difficult to handle.

Depending on work force requirements, construction camps not only may be very large but they may also require fairly elaborate installations. They usually need heating systems and their own power plant and generating facility for electricity and they may even require water treatment and sewage disposal plant facilities. Sometimes the camps become miniature towns in which there is a transient population of several thousand men. Under special circumstances quarters are set up for married couples but they are the exception rather than the rule.

At the end of the project it is customary to sell the camp facilities for salvage, although they are sometimes taken over by the plant owner or management to provide accommodations for future employees until permanent facilities are available.

STATISTICS

Compiling statistics on a project can prove of great value to owners, consultants, and contractors alike. In addition to providing individual project data for instant recall, statistics can be used for evaluating other projects or for indicating trends which can be considered in future actions. It is easy and inexpensive to compile statistics for most buildings, and by using an appropriate format the data can be processed by a computer for future use.

On the following pages are examples of relevant statistics.

Checklist

BUILDING STATISTICS

No:
Date:
Project:
Location:
Type of building:
Owner:
Architect:
Engineers: Structural:
                Mechanical:
                Electrical:
Other Consultants:
Services Supplied:
Tender Date: Amount: $
Contract Award Date: Amount: $
Extras/Credits: Amount: $
Completion Dates: Contract: Actual:
Final Cost: $

Area:	Sq ft (m^2)	Unit Cost: $	/Sq ft (m^2)
Volume:	Cu ft (m^3)	Unit Cost: $	/Cu ft (m^3)
Structure:		$	/Sq ft (m^2)
Plumbing — Heating — Ventilation:		$	/Sq ft (m^2)
Air Conditioning:		$	/Sq ft (m^2)
Electrical:		$	/Sq ft (m^2)
Foundations: Piles:	Footings:		Raft:
Reinforcing Steel:	lb/Sq ft (kg/m^2)		
Structural Steel:	lb/Sq ft (kg/m^2)		
A.C. Tonnage:	Sq ft/ton (m^2/t)		
Special Conditions:			

Checklist

MISCELLANEOUS FORMS USED BY GENERAL CONTRACTORS

(Note: Forms use Contractor's Letterhead)

Cost Estimate
Project:
Date:

Code	Description	Quantity	Unit Prices L M E S	Extensions L M E S	Total

L = Labor
M = Material
E = Equipment
S = Subtrades

Signed:

Cost Report
Project:
Date: Period covered:

Cost code	Estimated	Actual	Revenue	Variance

Signed:

Job Meeting Minutes
Project:
J. M. No.:   Date:

Item No.	Description	Action by:

A. Repeat items
B. New agenda

Signed:

MISCELLANEOUS FORMS USED BY GENERAL CONTRACTORS
(continued)

Daily Report
Project: Date:
                                      Weather:
                                      Temperature:
List of Activities   Precipitation
Work force   Wind:
Equipment   General Contractor/Subtrades
Material received   General Contractor/Subtrades
Special visitors   General Contractor/Subtrades
Concrete placed   General Contractor/Subtrades
Steel erected
Special remarks
                                                        Signed:

Weekly Report
Summary of daily reports plus activities planned for next week
                                                        Signed:

Weekly Progress Report
Project: Week ending:
List of activities   Percent complete
                                      This week   Cumulative
                                                         Signed:

Concrete Placing Record
Project:
Date   Location   Strength   C.Y.
                                            This pour   Cumulative

---

Daily Time Card
Project: Date:
Name   No.   Work items
                                     Hours worked   Total
                                                    Signed:

MISCELLANEOUS FORMS USED BY GENERAL CONTRACTORS
(continued)

Reinforcing Steel Setting Record
Project:
Date   Location   Tonnage
                                                   This item   Cumulative

---

Progress Payment Request
Project: Date:
No.   Period Covered

Original contract amount:
Approved change orders: _____
Revised contract amount: ══════════════

Completed to date: %
Claimed this period: %
Less previously claimed: _____

Subtotal
Less holdback: %   _____

Total payable:

Back-up required: Detailed breakdown as per control estimate
                   Original contract
                   Change orders

                                                    Signed:

MISCELLANEOUS FORMS USED BY GENERAL CONTRACTORS
(continued)

Drawing Transmittal
Project:
Sent to: Date:
Drawing No.   Revision No.   Description   No. copies   Instructions

Signed:

Drawing Record
Project:
Drawing No.   Revision No.   Description   No. copies   Date received

---

Request for Change Order
Project:
No.   Date
Description

Addition/Deduction

Signed:
Approved on: Signature:

Change Order Record
Project:
No.   Description   Date   Date $ $ Remarks
                        Requested   approved   Cumulative

Job Instructions
Project:
No.:   Date: Ref.:
Instructions:

Authorized Signature:

Requisition
Project:
Buy: Rent: Lease:
Code No.   Description   Quantity   Unit   Cost   Delivery   Order No.
                                              price   date

Requested by:
Approved by:

MISCELLANEOUS FORMS USED BY GENERAL CONTRACTORS
(continued)

Purchase Order
Project:
No.:   Date:
Delivery Date: Location: Via:
Sales Taxes: Terms:
Code No.   Quantity   Description   Unit   Total
                                                Price

                                                        Signed:

Material Receipt
Project:
No.:   Date:
Requisition No.:   Purchase order No.:
Dated: Dated:
Received from: Via:
Material: Equipment:
Code No.   Quantity   Description   Unit   Total
                                                price

                                                        Signed:

CLAIMS

Hardly any contracts are ever completed without there being a number of changes resulting in extras and credits to the contract. The majority of these originate either with the contractor or with the consultants, and they may start as a "change notice" which eventually results in a "change order" to the contract. An addition to the contract may be adjusted on the basis of net cost plus a set percentage (a common contractual method) or it may be on a negotiated basis and/or a lump sum. These are the customary approaches in most cases, and as long as the amount of the changes remains within a reasonable percentage of the original contract, barring major changes to the project, there are seldom arguments or objections. If, however, these changes are of a substantial nature, the element of time enters the picture. The contractor who is faced with extras amounting to a substantial percentage of the contract may contend that it will take so much longer to complete the contract and will request an extension of time. In addition, there will probably be a claim for extra compensation for extended duration costs for the project.

There are many claims which arise from other causes and which may be of a contentious nature. They do not fall under the category of normal extras and are usually brought in by the general contractor. A number of such claims are outlined in the checklist at the end of this chapter.

Claims are often made over subsoil conditions that are not the same as those specified in the tender documents. Such claims are often combined with requests for extra compensation and additional completion time.

Subcontractors may submit claims to the general contractor. And the owner or his representatives may make claims or backcharges against the general contractor or his subcontractors for damage caused, work not performed according to the contract, delays in project completion, or other grievances.

When modern planning techniques and methods are used, claims can evolve strictly as a result of sharp analysis of certain events and progress/time relationships. It is possible for a contractor to complete a building on time or even earlier than required by the contract and yet be delayed by the owner under certain conditions that would provide grounds for a delay claim and potential compensation.

When network techniques such as CPM and similar methods are used, arguments might arise over the disposal and rights to "float" time. Many people tend to look at float time as a convenient surplus, but more sophisticated contractors look at it as a valuable resource that enables them at times to change their approach to a certain project or parts thereof.

All claims, of course, must be related to the contractual considerations and must be evaluated accordingly. Since most claims tend to be of a contentious nature, it is often advisable to use independent and unbiased parties to assess them. Obviously, the contractor or owner, depending on who originates the claim, requires an evaluator. The consultants, who are usually the first to be questioned, tend to be biased because very often the nature of the claims directly or indirectly reflects on their professional ability and conduct. The same reasoning may apply to an owner's project manager.

If a claim is to find good reception, it must be properly presented, clear and to the point, concise but not too abbreviated, and well supported by substantiating evidence and proofs. Supporting diagrams often lend valuable assistance to the presentation.

A chronology of events must be presented if the claim is a delay claim. There an "as-built" detailed schedule is essential, especially if the causative factors can be demonstrated graphically. The various claim items should be dealt with on an individual basis and then summarized in corollary.

CLAIMS ASSESSMENT

The first step in analyzing a claim is to determine whether the claim is contractually or legally justified and valid, that is, to determine the entitlement. This requires great care and discernment because the initial impression can sometimes be deceptive and often only an in-depth study may reveal the true nature of the claim.

If the claim has been examined and found invalid, the matter stops right there. If the claim has been found to be a valid one, then the next step involves evaluation of the quantum (the amount of entitlement or compensation the claimant is due). This may be equal to but it could be more or less than the amount claimed.

Even though a claimant knows that he has neither contractual nor legal grounds, he may nevertheless present a claim on a *compassionate* basis. This presupposes a certain amount of generosity on the part of the recipient of the claim. More often than not the claim falls on deaf ears, but sometimes the other party finds some self-serving interest in considering the exercise of compassion.

Depending on the nature of the claim item, it may warrant extension of time without financial compensation, or financial compensation, or both. Contractual terms will usually determine the appropriate choice. Examples of claims which may warrant extension of time, but not necessarily financial compensation include: Force Majeure (Acts of God); wars or civil uprisings; new state laws; natural disasters; strikes or lockouts, or other labor problems; unusual weather conditions; other contingencies specified in the contract.

The assessment of a claim depends mostly on the quality of the back-up material presented. The better and more complete the documentation, the easier it is to assess the claim. Many a claimant has lost out on a perfectly valid claim because he or she failed to present proper substantiation.

General contractors are notorious for neglecting to maintain proper project records. The importance of keeping a detailed daily log cannot be overemphasized. The log should not only list all activities, but it should also chronicle both the start and the end of specific operations because they may eventually become prima facie evidence in a claim in which the time factor plays an important part, such as in a delay claim. If the case were to end up in court, this evidence would become even more important.

An "as-built" diagram is often useful in evaluating and it should be constructed from the information derived from the various sources used to make the evaluation.

Checklist

CLAIM ITEMS

Delays
- Late Contract Award
- Late Issue of Drawings
- Late Approval of Shop Drawings
- Hold on Partial or Complete Contract
- Lack of Information
- Late Decisions
- Late Supply of Owner-Supplied Material
- Late Completion by Others
- Late Delivery of Work Areas
- Interference by Owner
- Interference by Others
- Excessive Additional Work
- Excessive Changes
- Complex Changes
- Increases in Cost of Labor
- Increases in Cost of Materials
- Increases in Equipment Rentals
- Extended Site Supervision
- Extended Site Overhead
- Contract Extending into Winter
- Snow Removal
- Dewatering
- Strikes
- Labor Slowdowns
- Loss of Revenue
- Escalation
- Additional Bonding Charges

Additional Costs
- Additional Work Orders
- Difference in Subsoil Conditions
- Drawing Revisions
- Additional Tests
- Travel Allowance
- Overtime
- Accommodations
- Winter Conditions
- Temporary Heat
- Loss of Productivity
- Conflict in Information
- Claims by Subcontractors

Special Claims
- Part of Contract not Paid
- Work Done under Protest
- Work Approved, not Paid
- Work Approved, No Change Order Issued
- Contested Backcharges
- Union Demands
- Contractual Losses
- Subcontractors Claims
- Future Claims
- Financing and Interest Charges
- Cost of Claim Preparation

CLAIM FORMAT

Letter of Transmittal	Back-up Information (By Items)
Summary of Claim	Schedules
Chronology (Delays)	Statistics
Justification (By Items)	Appendices (As Required)

Checklist

ASSESSMENT FORMAT

Introduction	Chronology
Summary	Source Documents
Assessment and Comments	Statistics
Recommendations	Schedules
Detailed Evaluation	Appendices (As Required)

Checklist

SOURCE DOCUMENTS

Claims	Estimates
Contract	Work Sheets
Tender	Time Sheets
Specifications	Payroll Records
Drawings	Job Ledgers
Correspondence	Cost Records
Progress Claims	Subcontracts
Payment Records	Purchase Orders
Change Notices	Invoices
Change Orders	Check Books
Work Orders	Labor Distribution Records
Field Instructions	Material Distribution Records
Job Reports (Daily, Weekly)	Equipment Rental Records
Job Meeting Minutes	Tool Records
Progress Reports	Expense Account Vouchers
Progress Schedules/CPM/Bar Charts	Labor Decrees
As-Built Schedules	Meetings and Discussions — Notes
Drawing Transmittal Records	

5

Subcontracts

In earlier times a general contractor did most of the work with his own forces, but today the majority of work is given out to subtraders specializing in either manufacturing, or installation, or supply and installation of various building components. In fact, the general contractor of today, with comparatively few exceptions, has essentially become a broker for the building trades. His functions are primarily the coordination and control of actual construction as well as of administration.

The scope and services of subcontractors vary greatly, and some firms combine a number of services handled individually by other firms as their specialty. Although any attempt at classification is likely to involve a certain amount of overlap, subtrades can be divided roughly along the lines given below.

SUBTRADES CONNECTED WITH SITE PREPARATION
AND FOUNDATIONS OF BUILDINGS

1. Demolition contractors
2. Excavation contractors
3. Utilities contractors
4. Foundation and piling contractors
5. Pressure grouting contractors
6. Landscape contractors
7. Paving contractors

8. Fencing contractors
9. Soil test laboratories

SUBTRADES CONNECTED WITH STRUCTURAL COMPONENTS OF BUILDINGS

1. Formwork contractors
2. Reinforcing steel fabricators
3. Reinforcing steel setters
4. Concrete placing contractors
5. Cement finishers
6. Structural steel fabricators
7. Structural steel erectors
8. Steel deck contractors
9. Asbestos deck contractors
10. Precast concrete contractors
11. Prestressing contractors
12. Slipform contractors
13. Lift slab contractors
14. Laminated wood contractors
15. Pneumatic concrete contractors
16. Specialty contractors
17. Concrete and steel testing laboratories

SUBTRADES CONNECTED WITH ARCHITECTURAL COMPONENTS OF BUILDINGS

1. Masonry contractors
2. Carpentry contractors
3. Ornamental concrete contractors
4. Curtain wall contractors
5. Metal siding contractors
6. Asbestos siding contractors
7. Cut stone and granite cutters
8. Artificial stone fabricators

9. Roofing and sheet metal contractors
10. Skylight manufacturers
11. Waterproofing and dampproofing contractors
12. Caulking and weatherstripping contractors
13. Insulation contractors
14. Metal window manufacturers
15. Metal door and frame manufacturers
16. Storefront and glazing contractors
17. Overhead and vertical lift doors contractors
18. Rolling door and grill contractors
19. Sliding door contractors
20. Flexible door contractors
21. Soundproofing contractors
22. Fireproofing contractors
23. Miscellaneous and ornamental metals contractors
24. Toilet stall and locker manufacturers
25. Vault door manufacturers
26. Folding door and partition manufacturers
27. Prefabricated partition manufacturers
28. Millwork manufacturers
29. Wire mesh partition contractors
30. Sign contractors
31. Hardware manufacturers
32. Awning contractors
33. Window washing equipment contractors
34. Specialty items contractors

SUBTRADES CONNECTED WITH ARCHITECTURAL FINISHES IN BUILDINGS

1. Plaster and drywall contractors
2. Painting and decorating contractors
3. Plastic finish applicators
4. Tilework contractors
5. Resilient flooring contractors

6. Mastic flooring contractors
7. Wood floor layers
8. Acoustic tile and ceiling contractors
9. Carpet layers
10. Blinds and drapes installers
11. Sculptors and artists

SUBTRADES CONNECTED WITH MECHANICAL AND ELECTRICAL SYSTEMS OF BUILDINGS

1. Plumbing and heating contractors
2. Air conditioning and ventilation contractors
3. Refrigeration contractors
4. Sprinkler and fire protection contractors
5. Electrical contractors
6. Chimney and incinerator contractors
7. Garbage compactor manufacturers
8. Lawn sprinkler system contractors
9. Supervisory and alarm system contractors
10. Lightning protection contractors
11. Music and public address system contractors
12. Cathodic protection system contractors
13. Restaurant and cafeteria equipment manufacturers
14. Kitchen equipment manufacturers
15. Laundry equipment manufacturers
16. Laboratory equipment manufacturers
17. Swimming pool contractors

SUBTRADES CONNECTED WITH VERTICAL TRANSPORTATION IN BUILDINGS

1. Elevator contractors
2. Escalator contractors
3. Moving ramps and sidewalks contractors
4. Automobile lift contractors

COMMENTS ON SUBTRADES

The following comments deal with individual trades and give suggestions on how to set up subcontracts and which specific points to mention or to look out for. It is important to note that regional and geographic differences exist for various building components and construction methods and that standards of workmanship and trade practices may also vary. These differences and standards must always be taken into consideration.

Plans officially released for construction may have errors, omissions, and ambiguities. The construction industry is used to awarding contracts and subcontracts on the basis of "as per plans and specifications." Faulty plans often result in disputes and/or claims, and eventually everybody loses out because the project most likely incurs delays and additional costs. The dispute may or may not be resolved, but the claims start to pile up. It is recommended strongly that the contract documents be specific and that the obligations of the contracting parties be listed in sufficient detail so that there will be no grounds for a dispute right from the date of signature. A few extra words in the contract can be worth a lot of time and money.

When fabricated items or materials are supplied, sales taxes are usually quoted separately. When items are supplied and installed, sales taxes are generally included, but this should be so stated in the contract documents because it could amount to a lot of money.

Certain items (tolerances, deflections of structural members, etc.) may be covered under specific professional or governmental codes or legislation or standards, and they should be referred to if they are relevant. Guarantees may be covered under legislation or industry standards and should either be stated specifically or else referred to.

It is often a good practice to list unit prices for adjustments of quantities on appropriate units in the contract. Prices negotiated at a later date usually have a tendency to be much higher since the subcontractor no longer has the competitive pressure he had at tender time.

It is of utmost importance to cover the timing. In the case of shop manufactured items, this timing may consist of the time to fabricate and deliver and the time to install. These are often tied in to the date of approval of shop drawings. If a schedule or CPM is used on the project, it may be incorporated in the contract documents, but often the information received from the subcontractors regarding timing of their own work becomes one of the major sources for the schedule.

Occasionally special incentives such as bonus or penalty clauses are built into the contract. Each subject must be treated specifically, depending on the prevailing conditions. Various special conditions to be allowed for in the subcontract are included in the checklist.

There may be other (prevailing) conditions which govern these clauses, e.g. the owner may want, if possible, the project completed earlier than the date specified in the tender. In this case, the general contractor may be given a bonus incentive and this may be echoed in some subcontracts which may be instrumental in helping to achieve the earlier completion.

When setting up references for the scope of the subcontract it is important to note that it may not be enough to base the subcontract on the appropriate section(s) only of the specifications but that special conditions, either typical for the project or relevant to the specific trade, must be allowed for and included in the subcontract.

Checklist

SUBCONTRACTS

The following is a list of the various items that may require inclusion in different subcontracts. The items to be selected depend on specific circumstances and are likely to vary in each case. Since it is impossible to cover all possible situations, it may be necessary to include additional clauses to cover specific conditions and circumstances.

The contracting parties are referred to as the "owner/contractor" and the "subcontractor."

Scope of work
  Work included
  Work specifically excluded
By subcontractor
  Items to be supplied
  Items to be installed/erected
  Items to be supplied and
    installed/erected
  Services to be supplied
By owner/contractor
  Items to be provided
  Services to be provided
Responsibility for temporary services
(by owner/contractor or by subcontractor)
  Heating and protection
  Lighting and power
  Snow removal
  Dewatering

Water supply
Sanitary facilities
Office space
Storage space
Telephone
Security
Clean-up
Responsibility for special services
  Shoring structure
  Providing access routes
  Layout, lines, and levels
  Preparation of work site
  Preparation of work surfaces
  Preparation of seating/supports
  Capping existing services
  Excavation and backfill
  Inspection and tests
  Hoisting materials/equipment

Checklist

SUBCONTRACTS *(continued)*

Cartage, unloading, handling
    materials/equipment
    Storing, protecting materials/
    equipment
    Bar lists, placing drawings
    Leaving forms in place (time)
    Leaving shoring in place (time)
    Insulating forms
    Heating concrete
    Cutting, patching
    Foundations, bases, supports
Schedule
    Duration
        Start dates
        Completion dates
    Phasing work
Labor and construction equipment
    Equipment to be used
        Type, capacity, size
        Number
        Transportation
        Travel time
        Storage
        Maintenance
        Operation
        Insurance
    Manpower
        Crew size
        Number
        Supervision
        Minimum workforce
        Working hours, overtime
        Restrictions
            Noise
            Dust
            Blasting
        Traveling time
        Board allowance
        Parking facilities
        Minimum/maximum working
           temperature
        Workmens compensation
        Unemployment insurance

Materials, equipment
    Quantity
    Quality
        New
        Used
        Standard
    Waste Allowance
    Paint/finish
        Prime coat
        Finish color
        Rustproof
        Waterproof
    Glue
        Waterproof
    Tolerances, clearances
    Fasteners
        Rustproof
    Ownership of salvaged materials
    Surplus materials
    Spare materials
    Samples
    Supplied F.O.B.?
    Time required
        Fabrication
        Delivery
    Shop drawings
    Models
    Guarantees, warranties
    Maintenance contract
    Replacing permanent materials
        used on temporary basis
    Restoring permanent equipment
        used on temporary basis
    Service manuals
    Approved substitutions
    Take-over conditions
    Electrical/mechanical
        characteristics
Payments
    Lump-sum
    Cost plus
    Fee
        Fixed

Checklist

SUBCONTRACTS *(continued)*

Percentage
Upset price
Savings participation
Penalty/bonus
Holdbacks
Discounts
  Trade
  Payment
Taxes
  Federal
  State
  Provincial
  Municipal
Permits, licences, royalties
Interest
Unit prices
  Extras
  Credits
Progress claims, applications
  for payments
Release of holdback
Final release
Definitions
  Pay lines
  Pay quantities
  Waste allowance
  Allowable costs
  Basis for changes
  Evaluation of changes
  Determination of
    Areas
    Volumes
Contract documents

Plans
Specifications
Applicable codes
Standards
Contract form
Insurance
  Fire
  Liability
  Property damage
  Automotive
  Glass
Bonds
  Performance
  Labor and material
Special site conditions
  Access
  Subsoil
Cooperation with others
Records to be kept and submitted
Subcontract
  Termination
  Cancellation
  Postponement
  Delay
Nonperformance
  Subcontractor
  Owner/contractor
Separate contracts by owner/contractor
Assignment of subcontract
Disputes
  Arbitration
Signatures, witnesses, date

6
Construction Equipment

Building construction, compared to heavy construction, is more material and labor intensive. Nevertheless, a great variety of equipment is utilized and the general trend pointing to a reduction in field labor tends to lead to ever-increasing equipment usage in building construction. There are a number of reasons for this tendency. Shop-produced building components are made under better controlled conditions, resulting in higher quality, and do not depend on the vagaries of climatic conditions. Labor problems can, and do, seriously upset project health. Labor productivity seems to be moving in inverse proportion to equipment efficiency. Although equipment costs represent a comparatively small proportion of building construction costs, ranging probably around 5%, their importance to the building is more indirect. Equipment helps to cut down on construction time and consequently reduces overhead costs.

In building construction most of the equipment may belong to subcontractors and the choice of what type of equipment to use may at times depend more on what is available rather than on what is best suited to the project. This could result in inefficiency and loss of time if left uncontrolled. Thus it may be advisable to cover approval of choice of equipment in the subcontract. There are a number of aspects to consider when choosing construction equipment including the following:

A. The purpose for which the equipment is to be used, that is, the type of operation to be performed and the nature of the material to be dealt with

B. The type and size of unit suitable for the above purpose, that

211

is, the capacity of the equipment, the strength of the engine, if motorized; restrictions, if any, of the work area; restrictions, if any, of access to and egress from the work area

C. The time allotted to the specific operations in which the equipment is to be used and the efficient scheduling of equipment usage.

D. Extraneous factors such as weather conditions either because of seasonal or climatic and/or geographic considerations, such as temperature gradient, wind pattern, maximum wind velocity, precipitation profile, and other relevant data

E. Availability of equipment suitable for intended purpose

F. Other specific site conditions or restrictions

There are other factors which are less important for individual building projects, for example, compatibility between different equipment items used on the same site and the ease of mobility of equipment on the site or between different sites. These factors determine whether equipment can be self-propelled or must be transported by carrier.

The effectiveness of the equipment ultimately depends on the skill of the operator. It may be necessary and, depending on circumstances, advisable, to conduct feasibility studies before making a final choice, especially in the case of major projects. When planning for equipment due consideration must be given to its removal from the site on completion of its task, for example, removing a tower crane from a high-rise building or removing an excavator from a deep, confined excavation.

Since equipment always performs functions dealing with materials, primarily handling and manipulating them, the equipment must be scheduled in conjunction with these specific materials to ensure its optimum usage. Since idle equipment does not earn revenue, it is particularly important to plan for maximum continuous use because it may not pay to remove the equipment temporarily and then bring it back.

A general contractor requiring equipment has the choice of three alternatives. These alternatives depend primarily on the time and use factor of the equipment which in turn may be a function of the nature of the project. The alternatives are:

A. Renting equipment

B. Leasing equipment

C. Buying equipment

Some of the considerations for each option are discussed below.

RENTING EQUIPMENT

Equipment is rented when it is required on a limited and short-term basis. The equipment is usually rented on "all-found" terms, which means that the equipment is fully operated and serviced and that all charges are included in the rental. Rental is on the basis of actual time plus an allowance for transportation or traveling time.

LEASING EQUIPMENT

Equipment is leased when it is required on a long-term basis. The equipment may be supplied by the lessor and operated by the lessee who becomes responsible for maintenance. Leasing would be feasible if the contractor were to envisage a long duration for equipment usage and renting would become uneconomic.

BUYING EQUIPMENT

Buying equipment is only worthwhile if the equipment is required on a long-term basis and if the contractor has reasons for wanting to assume the costs of owning and operating the equipment. The main difference between buying and leasing the equipment is the capital expenditure involved in buying equipment outright. For this reason, many equipment leases have clauses which allocate part of the lease payments towards purchase of the equipment after a certain time period.

Thus when the economic feasibility of owning equipment outright is being considered, the contractor must take into consideration the cost of ownership and the cost of operation.

The cost of ownership includes capital investment and interest, taxes, insurance, storage and idle time, periodic major overhauls and downtime, depreciation and reserve fund.

The cost of operation includes operator's wages, fuel and lubrication, tune-ups, maintenance and minor repairs, tire wear and repairs, assembly and dismantling, transportation of equipment.

The principal equipment items and related accessories in building construction include:

A. Excavation, backfill and sitework
  1. Power shovels and backhoes
  2. Dump trucks
  3. Dragline and clamshell excavators
  4. Earth augers
  5. Loader/backhoe units (pepines)
  6. Trenchers
  7. Front end loaders
  8. Bulldozers
  9. Graders
  10. Scrapers
  11. Rollers and compactors
  12. Tampers
  13. Compressors
  14. Crane and wrecker's ball
  15. Tractor-trailers
  16. Water trucks
  17. Soil-cement equipment
  18. Paving equipment
  19. Sweepers
B. Foundations and shoring
  1. Pile drivers
  2. Pile extractors
  3. Caisson drills
  4. Generators
  5. Welding machines
  6. Boilers
C. Structural work — steel and concrete
  1. Cranes
  2. Derricks
  3. Ready-mix trucks
  4. Concrete pumps
  5. Belt conveyors
  6. Vibrators
  7. Concrete buggies

8. Concrete finishers
9. Tremie and elephant trunk
10. Concrete bucket
11. Concrete saws
D. Temporary services and materials handling
1. Construction hoists
2. Tower cranes
3. Mobile cranes
4. Forklift trucks
5. Trucks
6. Heaters
7. Pumps
8. Snow blowers
9. Trailers
E. Miscellaneous building trades
1. Scaffolding, jacks
2. Special equipment items

In the following sections are brief descriptions of the various types of equipment listed above. Capacities are given in the building construction range. Accessories are stated separately when sufficiently important.

EXCAVATION, BACKFILL, AND SITEWORK

Power Shovels and Backhoes

The main distinction between power shovels and backhoes, which may use the same basic power unit but use different attachments, is that shovels work at or above grade elevation and backhoes work below grade. Grade here means the elevation of support of the equipment itself. Normally, grade refers to the elevation of the undisturbed surface of the construction site at any given point. It should also be noted that a shovel works with the bucket away and up from the cab, moving in a forward direction, whereas a backhoe works with the bucket down and toward the cab, moving backward and away from the excavation. Thus a shovel is used to cut into a pile of material and a backhoe is used to dig basements and trenches. The material excavated may be cast only, in which case the machine turns at right angles to the direction of travel, or the material is loaded on trucks and carted away, in which case the machine may turn at anywhere from 90° to 180°, depending on the optimum approach of the hauling equipment. Needless to say, the less the machine has to turn, the higher will be its output and effective usage (Figures 6–1 and 6–2).

In most building construction power shovels and backhoes are used in a range from ¾ cubic yard to about 4 cubic yard. They are generally mounted on caterpillar tracks and are cable operated. There are hydraulically operated models and machines with telescopic booms on the market, but their size is limited.

Dump Trucks

Dump trucks are used in excavation to dispose of the surplus material that has to be carted away. They are also used to bring backfill and granular material to the site, including earth, gravel, crushed

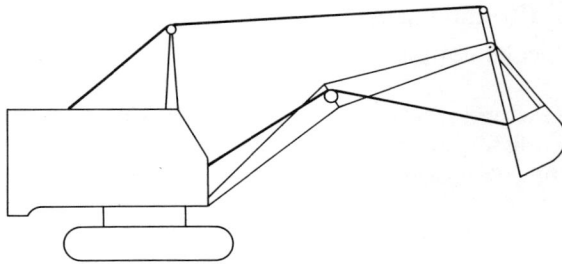

FIGURE 6-1. Backhoe.

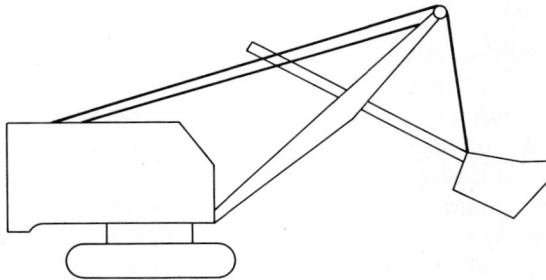

FIGURE 6-2. Shovel. Note: *basic power unit is the same. Attachments are different.*

stone, rock, etc. They come in an assortment of types and sizes and are generally limited by local maximum weight limitations on roads. They may range up to about 30 tons. Trucks larger than that or exceeding the weight restrictions are considered off-road trucks and are not allowed to travel on roads under their own power. They could be used on a building construction site, but they are designed for heavy construction sites such as mine developments, dam sites, etc. The size and output of the excavating equipment determine the number of trucks to allot to each machine. Too many trucks increase the waiting time between trips, but too few keep the excavating equipment waiting. Both of these situations must be avoided.

Dragline and Clamshell Excavators

These machines are rarely used in building construction. They are suitable in noncohesive or granular material or in wet soil, all of which are hard to handle with shovels or backhoes. With a dragline (Figure 6–3), the operator swings the bucket away from the machine and then pulls it toward the cab, an operation which requires considerable skill. With a clamshell, the load is handled vertically because it allows greater control in loading trucks. Bucket sizes may range up to 4 cubic yards.

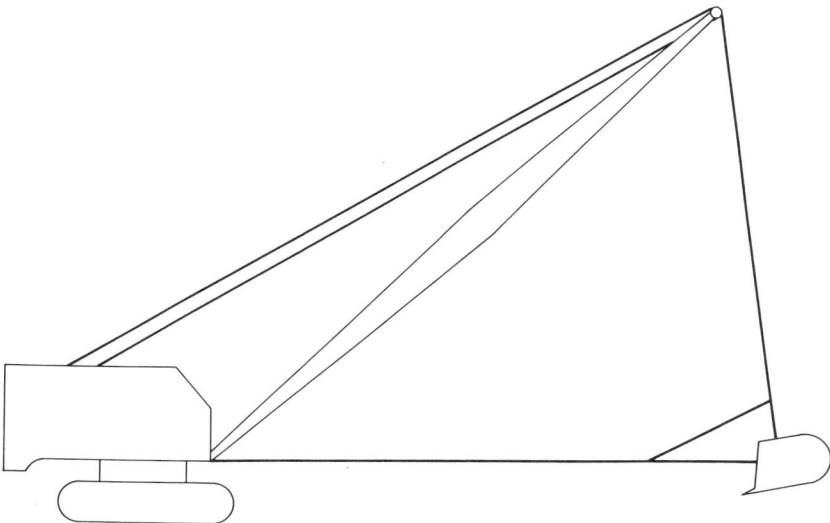

FIGURE 6-3. Dragline. Note: *Same basic power unit as backhoe or shovel. Attachments are different.*

Earth Augers

Earth augers are essentially large drills used to drill holes in compact and coherent soils. They are usually truck-mounted and come in a variety of sizes. The shaft may be sectional or have a telescopic arrangement. Some special machines are capable of drilling holes up to 10 ft in diameter and over 100 ft deep, but in building construction where augers are used more for implanting poles, the machines go up to about 3 ft in diameter.

Loader/Backhoe Units (Pepines)

A pepine is a smaller rubber-tired combination front end loader and backhoe. The machine has a bucket capacity at either end ranging between ½ cubic yard and 2 cubic yards and it is often used for excavating and backfilling trenches for footings and drains, pits, and other smaller jobs. Since it is highly mobile, it replaces the standard machines in locations which for reasons of access or capacity required preclude their use. When the machine is working, outriggers are extended laterally to give the machine stability and take the work load off the wheels and tires (Figure 6–4).

Trenchers

Trenchers dig narrow trenches by means of a bucket wheel. They may be used for grade beams and drain trenches, but their limitations

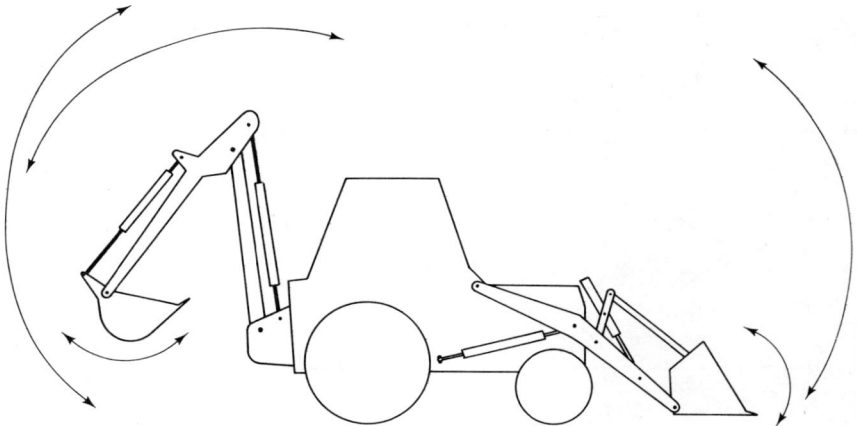

FIGURE 6-4. Loader/backhoe (pepine) showing composite motions of bucket.

do not usually make them suitable on building sites except perhaps on services for a large parking area or a major plant (Figure 6-5).

Front End Loaders

Front end loaders come equipped either with caterpillar treads or rubber tires and have buckets ranging in size between 1 cubic yard and 6 cubic yards, but there is an economic limit of the size to be used on building sites. They can be used for excavation purposes, but they are more suitable for loading surplus materials at grade elevation on trucks or for spreading granular fill under slabs. They are also used to level the bottom of an excavation or to spread top soil for finish landscaping work (Figures 6-6 and 6-7).

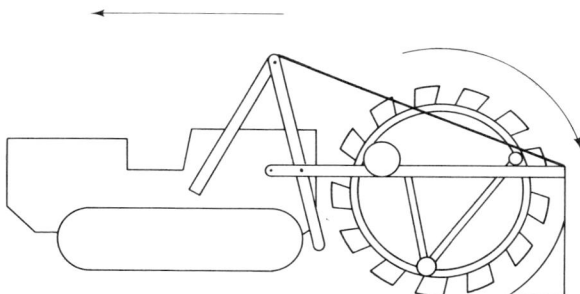

FIGURE 6-5. Trencher. Showing motion of bucket wheel and direction of travel.

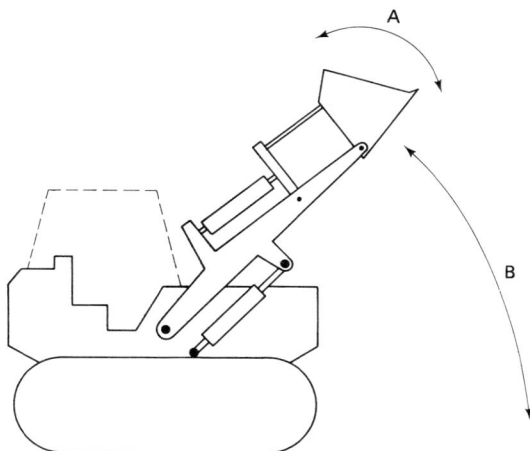

FIGURE 6-6. Front end loader (caterpillar tracks).

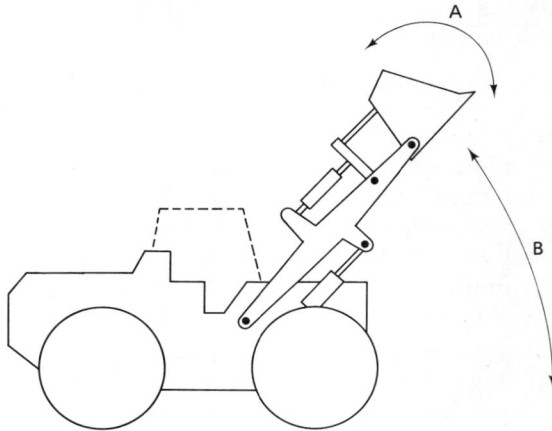

A: BUCKET TILTS
B: BUCKET IS RAISED

FIGURE 6-7. Front end loader (rubber-tired).

Bulldozers

Bulldozers usually come equipped with caterpillar tracks with or without lugs, or they may come in a rubber-tired version. The latter is found more in special models used for specialty applications and it generally uses very large units which are not found on building sites (Figure 6-8).

The blade size must be suited to the weight and power of the engine and tractor, but it may also vary depending on the purpose the machine is used for. The bulldozer may be equipped with special attachments: for example, rippers for breaking up tough soil, broken rock, frozen ground, or winches.

Bulldozers are used for cutting top soil, rough or finish grading, spreading materials, or as pushers for graders. Their capacity is usually expressed by their weight, but the blade size is also sometimes used to express capacity.

Graders

Unless there is a large parking area or extensive roadwork involved, graders are not generally used on building sites. Sometimes they are used to clear snow (Figure 6-9).

They may be rated by engine power and the blade size must be proportioned to this. They are used to level surfaces and shape

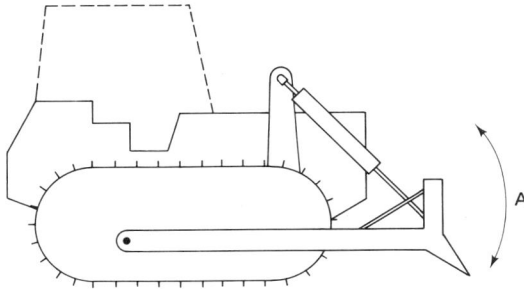

A: MOTION OF BLADE

FIGURE 6–8. Bulldozer.

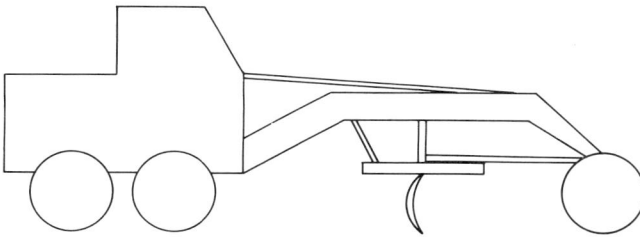

FIGURE 6–9. Grader. Blade can be turned and raised.

embankments because the blade can be turned and tilted and the wheels can move out of the vertical plane. Graders may be equipped with special attachments, such as a dozer blade for relatively light work, a snow plow with or without wings, or a scarifier for surface treatment.

Scrapers

Unless the site is very large and involves a substantial amount of earthmoving, scrapers are not likely to be used on a building site because their economy is geared to large volumes. They are used primarily in removing or relocating soil. The nature of the site and the soil composition determine the type and size of machine to be used. This choice of equipment is also a function of the distance traveled to the spoil or pick-up area.

Scrapers come in a variety of types and sizes, for example, crawler-tractor drawn, tire-tractor drawn, self-loading, elevating, etc. Scrapers may be assisted in loading by bulldozers that act as pushers. As a rule, one dozer serves several scrapers (Figure 6–10).

Scrapers are usually rated by volume capacity but at times the engine power is used instead. They may range up to 50 cubic yards heaped volume, but the capacity depends greatly on the material and the type of scraper since the soil picked up in the bowl becomes compacted and is denser.

The scraper consists of a bowl that is equipped with a front cutting edge, an apron that can be raised, and an ejector in the rear that is used to empty the bowl when discharging the material. First, the bowl is lowered to the ground. When the front apron is raised, the cutting edge cuts into the soil. When the bowl is full, the apron is lowered to close the bowl. To discharge the material, the bowl is opened again and the ejector is activated.

On dry, firm soil rubber-tired scrapers are faster and more efficient. As a rule, caterpillar-type scrapers are used in wet ground where rubber tires would not have enough traction or support. The cohesion and angle of repose of the scraped material influence the amount scooped up and the heaped volume of the bowl.

Rollers and Compactors

On construction sites compaction is required in some areas, for example, the subgrade under slabs on grade, granular fill under floors, backfill, and work in connection with paving and landscaping. For large areas, tractor-drawn or self-propelled rollers or compactors are used. They are usually rated by weight. Rollers may come as drum rollers, usually either in tandem or three-axle tandem (Figures 6–11 and 6–12). For added weight, the rollers may be ballasted with

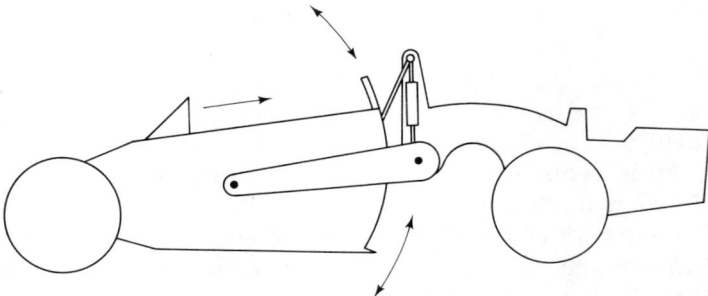

FIGURE 6-10. Scraper. Showing motions of bowl, front apron, and ejector.

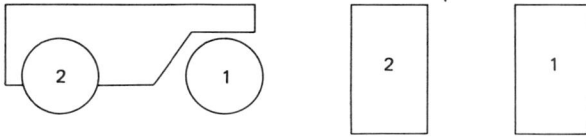

FIGURE 6-11. Tandem roller.

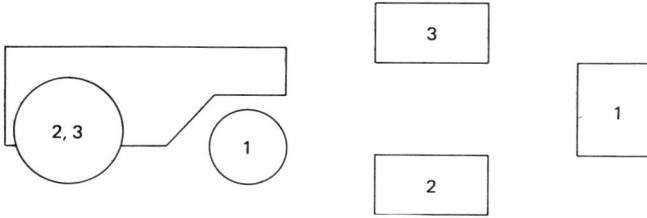

FIGURE 6-12. Three-axle tandem roller.

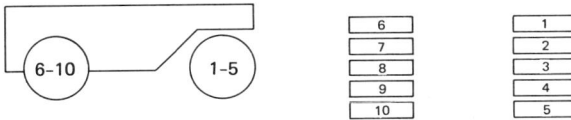

FIGURE 6-13. Rubber-tired roller.

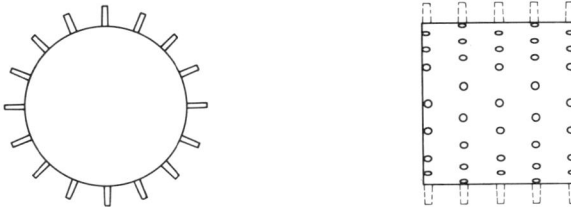

FIGURE 6-14. Sheepsfoot roller.

water. They may also come with pneumatic tires or as sheepsfoot rollers (Figures 6-13 and 6-14) for special treatment of subsoil that requires working action as well as compaction. They may be equipped with a vibrating mechanism or they may achieve a special effect by using wobbly-wheel rollers whose wheels are mounted eccentrically.

Tampers are used where the area is too small for wheeled equipment. They are motor-operated and achieve compaction by pounding or vibrating the material. They are especially suitable for exterior backfill, around footings and drains, at walls, and in other hard-to-get-at places. They may be rated by the energy output or by the number of vibrations per time unit.

Compressors

Compressors are used to generate pneumatic pressure for various purposes and to operate air tools. They are rated by capacity, usually in cubic feet per minute (cfm). As a rough indication of capacity, a paving breaker may require 50 cfm. Air tools include, among others, pneumatic drills, jackhammers, paving breakers, air tracks, wagon drills, clay spades, tampers, and vibrators.

Crane and Wrecker's Ball

A crane and wrecker's ball is used in demolishing existing masonry and concrete structures. The balls are rated by weight and vary generally between 1 and 3 tons. The crane required depends on the weight of the ball and the height of the structure to be demolished.

Tractor-Trailers

Many major pieces of construction equipment, mainly those used in earthwork, must be transported to the site because even if they could travel on their own, they are not allowed to use the roads. They are transported on large flat-bed or low-bed floats pulled by tractor units. They are rated by capacity and the range extends to about 50 tons, with limitations governed by local road restrictions.

Water Trucks

Water trucks are generally found on large sites only. They are used to keep down the dust by sprinkling water or they are used to supply water on sites having no temporary water facilities. They are rated by capacity and range from 1000 to 5000 gallons. The water may be dispensed by gravity or with the aid of pumps.

Soil-cement Equipment

On some major paving sites, such as parking lots for regional-sized shopping centers or large plants, soil-cement may be used for the parking area instead of conventional paving. Soil-cement requires specialized equipment, for example, pulverizers, cement spreaders, etc., which are either towed or self-propelled. The use of soil-cement, which consists of a mixture of cement and native soil, may be limited by climatic or geographic conditions. It can, of course, only be considered if the existing soil is of a suitable nature and consistency.

Paving Equipment

Paving machines are used to finish the bituminous surfaces on parking areas and roads which are mostly asphalt paving. Pavers are rated by their width and may range between 6 ft and 12 ft. The capacity and amount of paving laid depend on the thickness of the courses. Ancillary equipment used may include heaters, tar kettles, tanks, etc.

Concrete paving machines are seldom used in building construction unless there is extensive concrete roadwork. They usually run on tracks and are rated according to their width.

Batching plants are not usually used on building construction sites and are considered outside the scope of this book.

Sweepers

After most construction operations are finished, it is sometimes expedient to use a sweeper to clean up. Sweepers range from small walking or riding models, often used in shopping centers or larger office buildings, to self-propelled road models which contain a large water reservoir and have high maneuverability. The driver is often positioned on the curb side so that he can gauge the position of the machine relative to the curb.

FOUNDATIONS AND SHORING

Pile Drivers

Pile drivers are used to drive piles required for the support of the building or for shoring or sheet piling. Depending on the nature of the operation and type of piles, leads may be used to guide the piles and hold them in position. They are absolutely essential for batter piles which are implanted at an angle to the plumb line.

There are several types of pile drivers, including the following:

A. Drop hammers, which are simply weights that work by gravity

B. Single - and double-acting air or steam hammers (and modified versions of both), which have additional impact force from the pressure generated in the cylinder

C. Vibratory pile drivers, which lower the pile by destroying the skin friction through vibration but depend on the nature of the soil the piles are driven through

D. Diesel hammers, which after starting up continue automatically by lifting the pile hammer by means of the recoil from exploding diesel fuel; the action is stopped by regulating the fuel supply.

Pile drivers are rated by energy output, but they may also be classified in terms of hammer weight. Leads may also have to be used when piles have to be positioned accurately, for example, in a pile cluster.

Pile Extractors

Some pile drivers can be reversed to act as extractors, but there are special air or steam driven extractors as well as some that work by vibratory action. They may be needed to extract piles which are not required or have been implanted incorrectly.

Caisson Drills

Caisson drills are used to core out the inside of caissons after they have been driven or to assist sinking the caissons into place. They are essentially large augers.

Generators

Generators are used to generate electricity for temporary power and lighting, and they may be required for certain vibratory pile drivers. They may be either air-cooled or water-cooled and are rated in a range extending from about 1 kW to 1000 kW.

Welding Machines

Welding machines are required to generate the current used for welding sections of piles or equipment or the walers and diagonals for shoring an excavation. They are rated by their electrical capacity.

Boilers

Unless they are already equipped with a boiler, steam-driven pile drivers require a boiler. The boiler may be mounted on skids for greater mobility and is rated by capacity ranging anywhere up to 150 BHP (brake horsepower).

STRUCTURAL WORK – STEEL AND CONCRETE

Cranes

There are many different types of cranes, but here reference is made specifically to the large mobile cranes which may be either truck-mounted or on caterpillar tracks. They are used for many purposes in construction either for lifting construction materials or building components or for handling other equipment. In excavation they are used to handle dragline or clamshell buckets to excavate under special site conditions. They are used in the erection of structural steel, in hoisting mechanical equipment to roofs of buildings, and in demolishing existing structures with wrecker's balls. In concrete work they are used to place concrete by bucket. They are even used to remove other cranes from buildings or to extract excavation equipment from deep excavations.

Cranes are generally rated by capacity, which must be equated to the radius and height of the load relative to the position of the crane; they may range up to about 300 tons. In order to increase stability, outriggers must be used and positioned firmly before lifting operations begin to prevent the crane from toppling over.

Larger cranes are usually cable operated; smaller cranes may be hydraulic or have telescopic booms. The effective reach may be assisted by adding a jib to the boom.

Derricks

Derricks are seldom used in building construction. They are sometimes favored by steel erectors because of their ease of erection and mobility (they move up with the frame as required). There are two main types: the stiff-leg derrick in which the boom is larger than the center mast and the guyed derrick in which the boom is lower than the center mast and is able to rotate around it. Derricks are rated by lifting capacity, which also becomes a function of the angle of the boom.

227

Ready-mix Trucks

Few building construction projects have enough concrete to warrant installation of a batch plant and few contractors bother to make their own concrete. Thus most concrete for buildings is ready-mix supplied in trucks that are owned by the supplier.

Ready-mix trucks are rated by capacity. The size limitation may be a function of road restrictions on local roads and highways. They range up to about 15 cubic yard capacity. When concrete is ordered, the discharge time must be kept in mind so that trucks are not kept waiting at the site for periods long enough to affect the quality of the concrete.

Concrete Pumps

Concrete pumps are used to place concrete by pumping it into place rather than by conventional means such as bucket or buggies which sometimes cannot be used because of the location of the pour. The pumps are either operated with compressed air or by pushing the concrete directly through a pipe to the location of the pour. Either system has a height limitation and the output is affected by the total length of the pipe and the number of turns in it. The pump capacity is usually rated in cubic yards per/hour for a specified length of pipe. The equipment must be cleaned carefully after every use so that concrete will not set on the walls and obstruct the flow in the pipe.

Belt Conveyors

Belt conveyors are sometimes used to place concrete in hard to reach places or directly from the truck. They may also be used to place backfill or granular material. Since the speed of conveyors is variable, they cannot be rated by capacity, but they may be classified by width of belt or by length.

Vibrators

Vibrators are required to ensure uniform placing and compaction of concrete in place. They may be driven electrically or pneumatically or have a motor built-in in their head. They are rated by the size of head or by the power output, but they may also be classified by the vibrations per time unit.

Concrete Buggies

Concrete buggies are used for placing concrete, but they may also be used for transporting other materials around the site, such as reinforcing steel or formwork material. They are generally pneumatic-tired hand-pushed carts with a tilting bucket, but they may also come in motorized versions, either for pushing or riding. They are rated by capacity and range up to about 2 cubic yards.

Concrete Finishers

Concrete finishers are motorized rotary trowels used by cement finishers to finish the surface of concrete floors. They are rated by blade size. It is usually necessary to implement them with a certain amount of hand labor, since their round configuration does not permit them to reach corners or obstructions in the floor.

Tremie and Elephant Trunk

Tremies and elephant trunks are used when concrete has to be placed in water or in deep forms. Because of their weight, especially when full of concrete, a crane must be used to support and position them.

Concrete Buckets

A concrete bucket is the most widely used means of moving concrete from the truck to its final location. It is used with a mobile or tower crane, or it is specially adapted to be used with a construction hoist combined with a hopper at the pouring floor. It is rated by capacity and may range from ½ cubic yard to about 4 cubic yards.

Concrete Saws

Concrete saws are used for saw cut joints in concrete slabs and for cutting openings. The saws may be self-propelled walking-type machines and are rated by power output.

TEMPORARY SERVICES AND MATERIALS HANDLING

Construction Hoists

Construction hoists are used for hoisting materials and placing concrete. They may even be used as temporary elevators, although this is not recommended and it is usually prohibited. They usually go up with the building and are, therefore, only effective once the building reaches a certain height. They also are limited in both the capacity and in the physical dimensions of the items to be hoisted. With the advent of tower cranes construction hoists are being used less and less on buildings. Smaller versions having simplified guide rails are used as material skips. The hoist consists of the tower which is made up of modular units, the winch and rope, and the platform or car. Intermediate platforms are usually extended to the tower from each floor. Since there is no such thing as a self-leveling device, the winch must be manually manipulated and the skill of the hoist operator is an important factor in the efficiency of the system.

Tower Cranes

For a number of years tower cranes have become extremely popular on the North American continent, but they have been used in Europe for many years. They may be the climbing type, that is, they move up with the building by jacking up special telescopic sections in the main shaft. They are often placed in the service or stair core which can be completed after the crane has been removed on completion of its work. They may be rated by capacity at a certain radius or by size of the boom and jib (Figure 6-15).

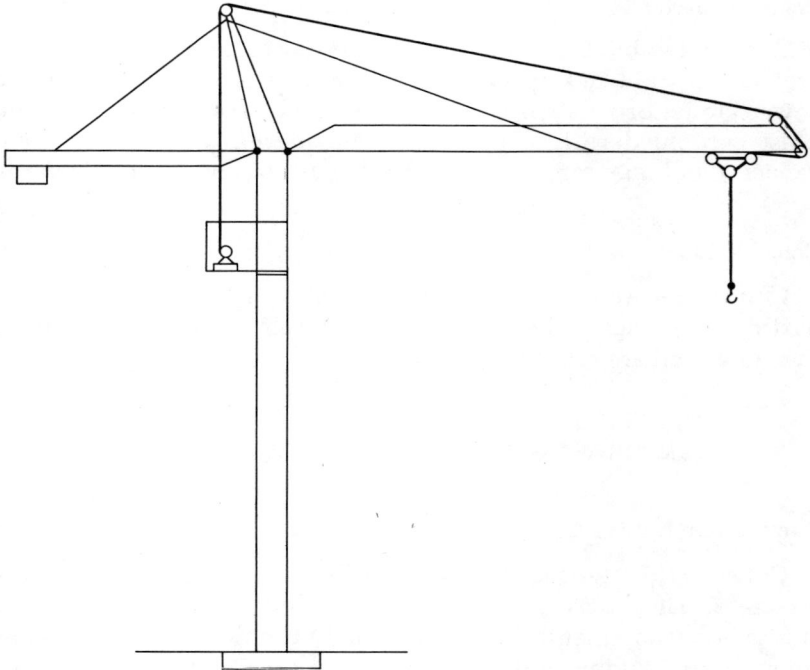

FIGURE 6-15. Tower crane.

Mobile Cranes

In addition to the larger mobile cranes described earlier, there are small mobile cranes that may be used for different chores on building sites. Some mobile cranes are small enough to be used inside the building. They are usually highly mobile and are rated by capacity (Figure 6-16).

Forklift Trucks

Forklift trucks are extremely useful on building construction sites for moving pallets or other materials. They are ideal in areas having restricted headroom and they can adapt themselves to the variable conditions often found on building sites, although they may sometimes be restricted by jacks shoring slabs or construction debris lying around. They are usually rated by their capacity.

Trucks

In addition to the trucks used for excavation, many other types of trucks are used on construction sites, for example, flat bed, stake,

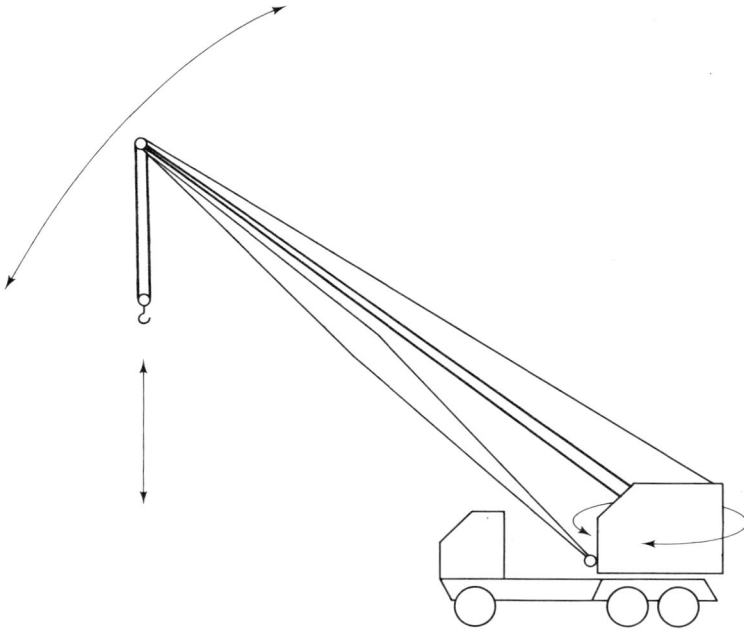

FIGURE 6-16. Mobile truck crane.

231

panel, pick-up, etc. They are used mainly for transportation of materials but pick-ups may also be used for transportation of supervisory personnel on the site. Each type of truck is rated by size or capacity.

Heaters

Buildings under construction often require temporary heating until the permanent system is in operation. Temporary heating systems may range from individual space heaters to complete installations.

Most space heaters used in building construction are either oil-fired or gas-fired (propane) and have a blower built in. They may also be ducted to distribute the hot air uniformly. They are rated by heat output and may range up to about 2,000,000 btu. It is extremely important to ensure proper ventilation for either type. Some fuels generate exhaust gases that contain not only carbon monoxide, which incidentally has adverse effects on concrete, but may also tend to deposit oily films if contained within the building. Other fuels, such as propane, generate large amounts of water vapor and can create moisture problems if not permitted to dissipate the vapor to the outside. Explosion hazardous fuel tanks, such as for propane, must be kept away far enough from the building so that there will be no danger. The fuel tanks may be subject to local codes in many areas.

Pumps

Pumps are used to dewater excavations using mud pumps, which can pump both water and suspended solids, and to remove storm water when the permanent drainage system is not yet in operation.

There are many different types of pumps, for example, centrifugal, reciprocating, diaphragm, pneumatic, submersible, etc. Pumps may be rated by the size of the discharge hose or by capacity, but these ratings must be equated with the head they work against and for which there is an upper limit.

Snow Blowers

Snow removal takes on many forms on construction sites, ranging from full use of large equipment on large sites to hand labor in confined areas. In locations falling in between and at locations inaccessible to mobile equipment it is handy to have a motorized walking-type or riding-type snow blower. The time and labor saved by using snow blowers over the duration of a project generally exceeds their cost by a comfortable margin.

Trailers

Trailers are often used for temporary field offices and storage areas. They are mobile and practical since they can be finished to suit specific site requirements and can be modified easily. Rubber-tired trailers should be jacked up and a skirt should be built around them to provide some insulation during winter and also to relieve the tires from their weight.

MISCELLANEOUS BUILDING TRADES

Scaffolding and Jacks

Many of the trades on building sites require scaffolding or jacks. These may be furnished by the trades themselves or by the general contractor. Most scaffolding is of a modular type and can be arranged to be installed either fixed in place or mobile by adding wheel attachments. Although many scaffold frames have built-in ladder rungs, many safety codes demand that separate ladders or stairs be supplied. Both scaffolds and jacks are rated by their load-carrying capacity.

Special Equipment

The various trades working on a building use a variety of equipment required by their work. This may include motorized drills, hammers, grinders, pipe threaders, saws, lathes, cutters, etc. They are mostly relatively minor and do not warrant more than a casual reference.

Index